William James Baldwin

Baldwin on Heating

Steam Heating for Buildings Revised

William James Baldwin

Baldwin on Heating
Steam Heating for Buildings Revised

ISBN/EAN: 9783743417113

Manufactured in Europe, USA, Canada, Australia, Japa

Cover: Foto ©berggeist007 / pixelio.de

Manufactured and distributed by brebook publishing software (www.brebook.com)

William James Baldwin

Baldwin on Heating

BALDWIN ON HEATING;

OR,

STEAM HEATING FOR BUILDINGS REVISED.

BEING A

DESCRIPTION OF STEAM HEATING APPARATUS FOR WARMING AND VENTILATING LARGE BUILDINGS AND PRIVATE HOUSES, WITH REMARKS ON STEAM, WATER, AND AIR, IN THEIR RELATION TO HEATING; TO WHICH ARE ADDED USEFUL MISCELLANEOUS TABLES.

BY

WILLIAM J. BALDWIN, M. Am. Soc. C. E.,
Member American Society of Mechanical Engineers.

With many Illustrations.

FOURTEENTH EDITION, REVISED AND ENLARGED.
FIRST THOUSAND.

NEW YORK
JOHN WILEY & SONS
LONDON: CHAPMAN & HALL,
1897.

THIS BOOK IS RESPECTFULLY DEDICATED

TO

MR. and MRS. WILLIAM DOUGLAS SLOANE,

IN CONSIDERATION OF

THE GREAT CHARITY THEIR BOUNTY HAS CREATED AND MAINTAINED

IN THE

SLOANE MATERNITY HOSPITAL, NEW YORK,

AND

THEIR CONCURRENT INTEREST IN THE SCIENCE OF

WARMING AND VENTILATION.

INTRODUCTION.

It was within the twenty years previous to the writing of the first edition of this book (1878) that the warming of buildings with steam carried through pipes became anything like a science; previously, it was a "chaotic mass of pipes, and principles," and even at the present time there is much blundering by persons who, being in some of the other engineering branches, imagine any one may do steam heating without any special training.

A low-pressure gravity apparatus is a healthful, fairly economical, and pretty perfect heating appliance, and may be constructed to heat a single room, or the largest building, with a uniformity which cannot be attained by any other means, except, perhaps, hot water.

By a gravity apparatus is meant, one without an outlet, whose circulation is perfect, wasting no water, and requiring no *mechanical means* to return the water to the boiler. It may with a hot water apparatus be likened to the circulation of the blood—the *boiler* being the heart; the *steam-pipes*, the arteries; and the *return-pipes*, the veins; thus carrying heat and life into every part of a building.

When reference is made to steam-pressure in this volume, it is understood to mean *pressure above the atmosphere*. Nearly all tables of reference on steam are given in *absolute* pressures—namely, pressures including the pressure of the atmosphere—which *unap-*

parent pressure has to be overcome before it is appreciable on a steam-gauge. As the steam-fitter has little, if anything, to do with pressures below atmosphere, the tables, etc., herein used will be modified, to commence at atmospheric pressure (14.7 pounds of the absolute scale), thus conveying comparison in the *ordinary terms* to which the steam-fitter is accustomed; and preventing the necessity of a mental calculation, which always involves fractions and enjoins a task which should not be thrown on a practical working-man or a beginner. Therefore, all pressures mentioned will be *apparent pressures*—namely, pressures which would be indicated by a *properly regulated steam-gauge.*

An endeavor will also be made to confine the work to facts as developed by practice, experience and experiment, refraining from comment and opinion except where the want of an explanation might lead to error or misunderstanding.

The first part of the book will be devoted to the principles of heating apparatus, etc., and the latter part to the details and to an amplification of such parts as are likely to change.

PREFACE TO THE FOURTEENTH EDITION.

In 1879 I wrote the first edition of this work, which truly could be called nothing more than a collection of hints, as stated in the preface of the earlier editions. By earlier editions I mean all editions published previous to this one. These editions, however, were the publisher's editions, being a reprint of each thousand books issued, with slight corrections, but without revision. So far as the work related to the principles of steam heating, where the water of condensation is returned by gravitation to the boiler, there could be very little change in the book, but to bring it down to modern practice in the use of steam by other methods, such as exists in large American cities, New York, Chicago and elsewhere, a general revision becomes necessary. It therefore became necessary that the whole former book be superseded by one whose date and practice harmonize. I will therefore endeavor once more to give some facts relating to the principles of modern steam fitting, which, since the writing of my first book, has made much progress, and I may say, has risen to the dignity of a branch of engineering science that may probably be known as domestic engineering, and which includes substantially all that goes to make up the engineering plant of a modern city building, excepting the electric light and elevator systems, which do not properly belong to it.

The order of the old book will be maintained as much as possible, and the chapters retained will be modified to conform with the present improvements in practice. The same, or nearly the same, title will be used, and the book in its old form will be withdrawn from the market.

THE AUTHOR.

CONTENTS.

CHAPTER		PAGE
I.	Gravity-circulating Apparatus	1
II.	Radiators and Heating Surfaces	24
III.	Classes of Radiation	37
IV.	Heating Surfaces of Boilers	50
V.	Boilers for House Heating	57
VI	Forms of Boilers Used in Heating	64
VII.	General Remarks on Boiler Setting and Construction	85
VIII.	Proportions of the Heating Surfaces of Boilers to the Radiating Surfaces of Buildings	93
IX.	Grates and Chimneys	99
X.	Safety Valves	114
XI.	Draft Regulators	129
XII.	Automatic Water Feeders	134
XIII.	Air Valves on Radiators	140
XIV.	Steam Pipe—Size, Area, Expansion, etc.	148
XV.	Size of Main Pipes for Low-pressure Steam-heating Apparatus, and Why Such Sizes are Necessary	163
XVI.	Steam	174
XVII.	Heat of Steam	182
XVIII.	Air	188
XIX.	High-pressure Steam Used Expansively in Pipes for Power and Heating	200
XX.	Exhaust Steam and Its Value	215
XXI.	Exhaust-steam Heating	223
XXII.	The Separation of Grease from Exhaust Steam	233
XXIII.	Boiling and Cooking by Steam, and Hints as to How the Apparatus should be Connected	242
XXIV.	Drying by Direct Steam	259
XXV.	Drying by Air Currents	269

CONTENTS.

CHAPTER		PAGE
XXVI.	Steam Traps	274
XXVII.	Valves for Radiators	285
XXVIII.	Remarks on Boiler Connections and Attachments	292
XXIX.	Data on Condensation in Radiators	298
XXX.	Pipe Covering—What is Saved Thereby, and Other Data	319
XXXI.	Miscellaneous Notes	325
XXXII.	Fire from Steam Boilers	331
XXXIII.	Miscellaneous Notes and Tables	337

CHAPTER I.

GRAVITY CIRCULATING APPARATUS.

WHAT is known as a low pressure gravity circulating apparatus is one in which the water of condensation from radiators and pipes returns to the boilers of its own weight without the assistance of mechanical contrivances. It is the apparatus used in warming private residences and oftentimes churches and schools. Churches and schools, however, at the present day generally have either electric lighting or fan engines and pumps, the condensed steam from which is of so much value that the gravity apparatus will probably soon be used only in the warming of private residences and detached buildings where steam for warming purposes alone is required. The term low pressure is a very general one, and usually it is accepted in the trade as meaning from atmospheric pressure to a pressure of about 10 pounds above it. An apparatus, however, that will work properly and return its water of condensation into a boiler at 10 pounds will work equally well at any higher pressure, and then it is simply called a gravity apparatus, and sometimes known as a high pressure gravity apparatus.

2 STEAM HEATING FOR BUILDINGS.

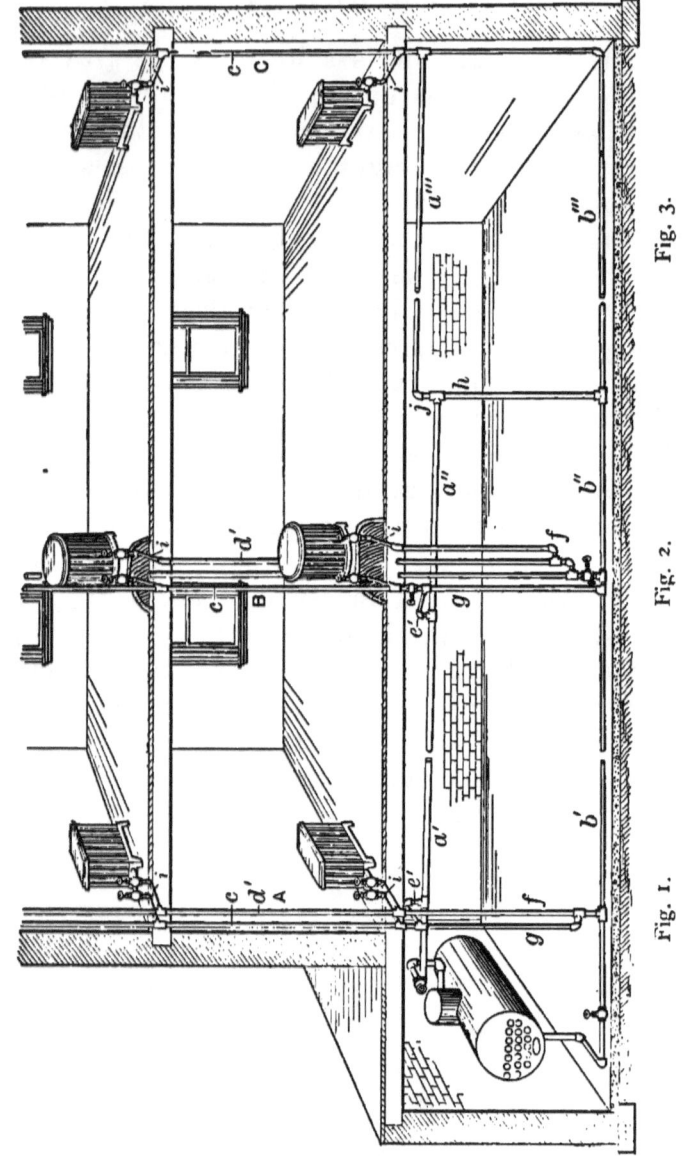

Fig. 1. Fig. 2. Fig. 3.

There are four systems of low pressure steam piping in use.

The first, and probably the best known, is what is to-day designated as the two-pipe system, in which the main return riser is carried below the water line. An example of it is shown in Fig. 1.

The second system employed is known as the separate return pipe system. It differs from the first in that the return pipe from every coil and radiator is carried below the water line of the boiler directly from the radiator before it joins and connects into the return pipe. It consists of the main horizontal distributing pipes, distributing risers, and a main horizontal return pipe, corresponding to the main distributing pipe, to which is connected the separate return risers from every coil or radiator; the returns not, connecting with each other until they go below the water line of the boiler. (See example in Fig. 2.)

The third system of gravity return consists of a main distributing pipe with distributing steam risers, including corresponding return mains, but no return water pipes from the radiators. The distributing riser carries the water of condensation back within itself to a relief pipe, carried below the water line, connecting with the main return pipe on the basement floor, which conveys the water into the return system, and thus to the boiler. This system is shown in Fig. 3. Fig. 4 shows a modification of this system, in which there may be said to be no return pipe. The steam pipe A starts from the boiler and runs all around the building of large diameter, gradually pitching downward to the point B, where it drops suddenly below the water line and enters the boiler at

C. This steam loop of large diameter around the whole basement becomes substantially a part of the boiler. The rising lines D D start almost directly upwards from it, and the water of condensation falls back within the same and goes on with the steam in

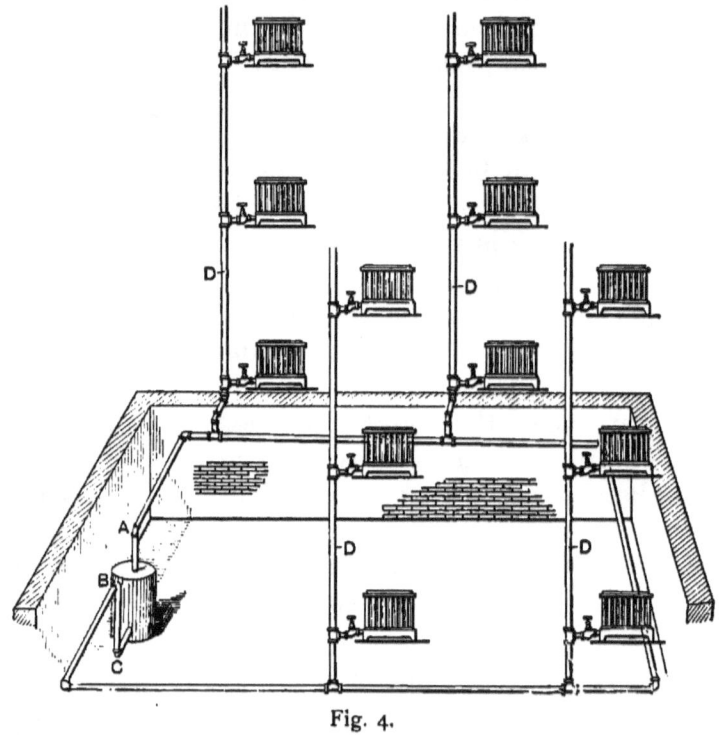

Fig. 4.

the direction of the arrows to the point C, where it enters the boiler.

There is a still simpler form of gravity return, and usually applied only to a very small building, which is shown in Fig. 5. In this case the diameters of the pipes in the basement are large and the horizontal distances short, and the steam pipe, when it leaves the head of the boiler, instead of dropping or pitching downward, runs upward at as steep an angle as possible. In this apparatus all the water of condensation from the radiators has not only to fall back within the steam rising line, but has to find its way back through the nearly horizontal main steam pipe into the boiler. If the pipes are sufficiently large in diameter the water will gravitate to the boiler in opposition to the steam flowing the other way.

System No. 1 can be run at *any* pressure, providing the pipes are sufficiently large in diameter and properly arranged; and it is the system commonly used in large buildings, both in the case of gravity apparatus and where exhaust and waste steam at very low pressures is used. It is well to add here that any of the system of piping just shown, excepting that of Fig. 5, which is purely a small house heating apparatus, will work under exhaust steam pressures slightly above atmosphere if the diameters of the pipes used are sufficiently large and slight modifications are made as mentioned elsewhere under exhaust steam systems. System No. 2 is generally *used in private houses*, and in buildings where extremely low pressure is employed, and with any of the first three systems can be made perfectly noiseless, when done with care, and there is rarely any difficulty in expelling the air when radiators are used.

For those not acquainted with the technical names of the different parts of a system, and to prevent misconception, the following explanation of terms is given. The same names always apply to the same part of the circulation, no matter what the system.

6　STEAM HEATING FOR BUILDINGS.

The word *circulation* here means the whole distribution of pipe in any one job or apparatus.

The Main Steam or Distributing Pipe.—The nearly horizontal live-steam main, generally near the cellar ceiling.

The Main Return Pipe.—The nearly horizontal pipe

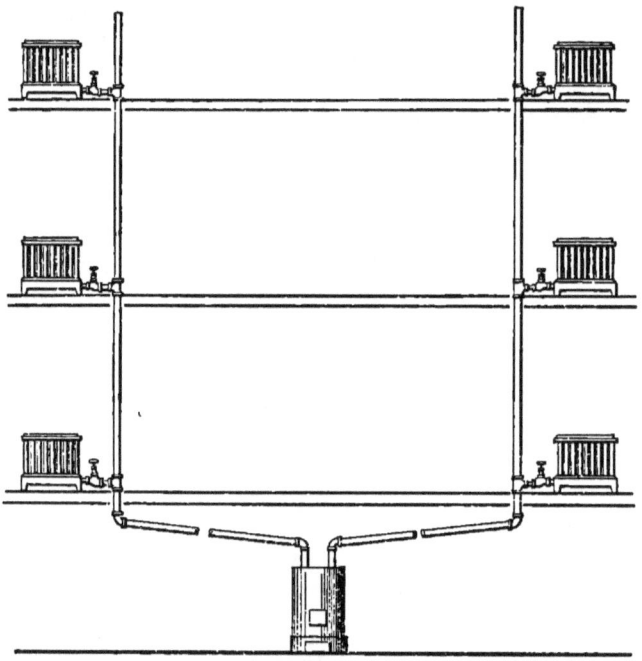

Fig. 5.

on the floor, or thereabouts, of the cellar, for carrying the condensed water back to the boiler.

The Steam Riser.—The vertical pipe which carries the steam from the main distributing pipe to the radiator connections.

The Return Riser.—The pipe which carries the condensed water from the radiators down to the main return pipes.

The Steam-Riser Connection.—The pipe which joins the main distributing pipe and steam riser.

The Return-Riser Connection.—The pipe which connects the return riser with the main return pipe on the floor.

The Steam-Riser Relief.—The pipe which connects the bottom of the steam riser with a T in the bottom of the return-riser connection or with the main return pipe, generally below the water-line. It carries the water which runs down the steam riser into the return-riser connection or main return pipe.

Main Relief Pipes.—Connections between the main steam and return pipes, to throw the water carried from the boiler, and that condensed in the main steam-pipe, into the return main, also employed as an equalizer of pressure in the system.

Radiator Connections.—The pipes which run from the risers to the radiators, both steam and return, usually no longer than is necessary to get sufficient spring for the expansion of the risers.

Rising Lines.—The steam and return risers taken together.

A Relay.—The jumping up of a main steam-pipe, with a main relief at the lower corner. This is to admit of keeping the main steam-pipe near the line of the risers and ceiling, and above the water-line, when the main lines are long.

Pitch—Is the inclination given to any pipe, and in the steam mains of a low-pressure or gravity apparatus, it should be down and away from the boiler so steam and water will always flow together (except in

System No. 5), and, if possible, the pitch should be toward the boiler in the main return. This is to have the water and the steam run in the same direction through the pipes, so as to prevent one source of noise in an apparatus.

Water-Line.—This is the term given to the general level of the water in the boiler and throughout the apparatus. In some cases, where the boilers are at a distance, or in a subcellar, and the fitter wishes to gain

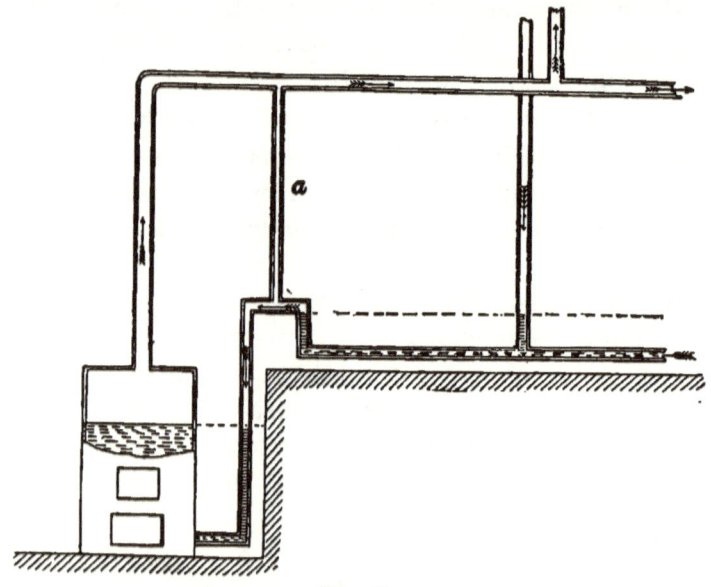

Fig. 6.

the advantages of having return mains and their return pipes and reliefs coming together *below water*, he makes *an artificial water-line* by raising the main return pipes higher than his connections before he drops to the boiler. It is also necessary to bring a relief, *a*, from the main steam-pipe to this *raised* part of the return to prevent siphoning into the boiler. Fig. 6 shows how this should be done.

It frequently happens in buildings where the line of the floor is below the water-line, that there are good reasons for not running the return pipe on the floor, when a modification of what is shown in Fig. 6 may be used; the return pipe being hung from the same hangers as the steam-main, and immediately below it, but raised about as shown before being dropped to the floor at the first available position. Still another

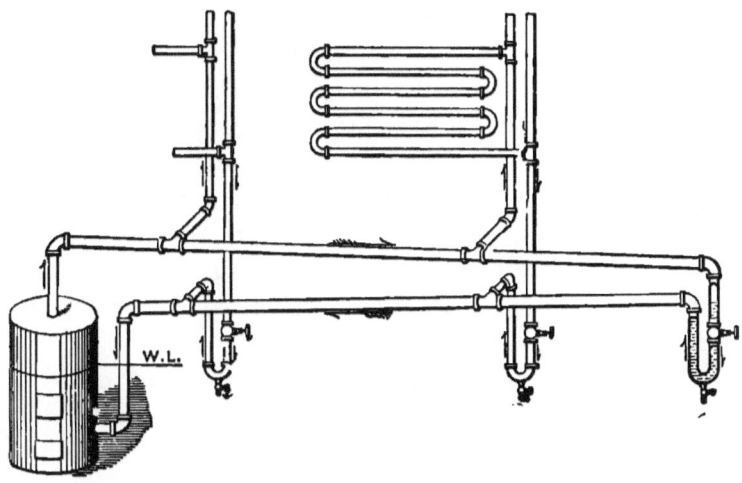

Fig. 7.

modification is to *trap* each return riser with an inverted water siphon by running the return riser some distance below the main return pipe, then rising and connecting with it, as shown in Fig. 7. When any of these means have to be resorted to, it would be well to have a pet-cock at their lowest points to draw the water from them in cold weather should they not be in constant use, as these water-traps may freeze.

HOW A BUILDING IS PIPED.

The steam-fitter should commence his work in a new building at an early period of its construction; and architects and parties paying for the work should see that the contract for steam heating be let when the mason and carpenter work is let.

The *risers* are the first work done in a new building constructed in the ordinary way. The architect should see that the builder and steam-fitter have an understanding at the commencement of the work and that the former leaves the proper recesses in the walls exactly where the steam-fitter requires them with proper regard to the strength of the walls. This will save much work to the fitter, and prevent the mutilation of the walls, and be no expense to the mason.

When the walls are up, the joists in their places, and the roof-boards or roof on, the steam-fitter should then put up his risers.

If the building has not more than three or four floors to be heated, it will answer to rest the risers on a support at the bottom of the recess; but in higher buildings the risers should be suspended by the *middle*, so that the expansion may be divided. By allowing the riser to go both up and down from the middle, the steam-fitter will be able to get along with shorter radiator connections, and will avoid the deep cutting of the floor joists in wooden floors.

The steam-fitter should avoid, as much as possible, taking many heaters from the same steam connection on a floor, and if it be unavoidable, he should if possible bring his return connections into the return riser some distance apart; or, he should run them *separately* down below the water-line, as it will prevent one heater from taking the air from the others or the return water from opposite directions meeting in the same fitting. When this cannot be done an enlargement of one size in the return piping when they come together should be provided so the opposing currents of water will not appreciably affect each other.

If the risers are on the side of the room, so that their outlets come between the joists, *it is best to keep the T's about half-way between the laths and the flooring,* as this admits of *nippling up,* and leaves room for *crossing* the pipes, if required, below the floor. But should the outlets come at the *side* of the joists, care must be taken that the T's come *in the exact place* to avoid the unnecessary cutting of beams. In a building with the risers resting on the bottom, and *all the expansion upward,* the top outlet must be the greatest distance below the top of the joist, but the top of the fitting must never come within less than ¾ of an inch of the floor when the riser is expanded to its utmost; so also with the rest of the T's, according to their distance from the bottom of the riser. The question of the expansion will be treated hereafter.

With low-pressure steam, the steam risers should be large. The general practice with steam-heaters is to reduce one size of pipe for each floor. This rule is not arbitrary; but as architects' specifications usually call for it there can be no objections, provided the piping is sufficiently large in diameter for its work.

In System No. 1 the return riser is generally one

size smaller than the steam riser, but it should never be smaller than $\frac{3}{4}$ of an inch pipe.

In System No 2, where many return risers are brought down in the same place, a 1-inch pipe for large heaters, and a $\frac{3}{4}$-inch pipe for small ones, are the sizes used generally. A $\frac{3}{4}$-inch pipe however will drain a 100 square foot heater if there is no great length of horizontal pipe.

When the risers are in place the outlets should be plugged up with *pieces of pipe* a foot or so in length, instead of the ordinary plug, as the latter is often *difficult* to get out when the recess is covered and the plastering done.

The risers should then be tested with cold water to 100 pounds per square inch. This will show if there are any cracked fittings or split pipe, and will save much time and annoyance when steam is gotten up.

When *automatic air-valves* are to be used on the steam-heaters, coils or radiators, a $\frac{3}{8}$-inch pipe should be run in the riser recess, with an outlet at each floor to receive the air-valve connection. The lower end of this air and vapor pipe should be taken to the nearest sink in the basement where any unusual waste of water or steam will be noticed by the engineer or janitor.

At this stage of the work, and before the floors are laid, the radiator connections should be run and firmly fastened in their places, making due allowance for the thickness of the floor, the furring on the walls, for the plastering, and for the baseboard. The radiator connections are usually run 1 inch or $1\frac{1}{4}$-inch for the steam connection, with a corresponding $\frac{3}{4}$ or 1 inch pipe for the return, according to the size of the heater; $1\frac{1}{4}$-inch steam-pipe being enough for a direct radiator of 150 square feet of heating surface, at low pressure.

When the radiators are threaded right-handed, the valves may be left-handed, to admit of connecting, by a *right-and-left-hand nipple* between the valve and radiator. When the valves are at opposite ends of the radiator, however, it is often difficult to use right and left nipples this way, unless there is considerable movement to the pipes under the floors, in which case either union valves must be used, or the right and left nipple must be between the lower end of the valve and the elbow under the floor.

When both valves are at the same end of the radiator, it is better to have the right and left nipples between the valves and the radiator. With this arangement both valves of the radiator can be connected simultaneously, and the movement of the radiator will be in the direction of the valves. It also admits of the disconnection of a heater after simply closing the radiator valves.

When the radiators are to be connected by any of the foregoing methods, the connections can be fastened (but not confined at their ends), so they may come in their exact places through the floors. The *free* ends of the connections should be closed with pieces of pipe long enough to come above the floors when the latter are laid. The air-pipe should also be run at the same time and brought through the floor in close proximity to the position the air-valve will occupy on the heater.

At this stage of the work the steam-fitter usually waits until the floors are laid, plastering done, partitions set and the basement graded.

Steam Mains.—Nearly all the success of the apparatus depends on its steam mains, *their sizes* and *how they are run.*

A heating apparatus has never yet been spoiled by having its steam mains large; still there should be a limit to their size, to prevent unnecessary expense and to keep the condensation and radiation of the distributing pipes at minimum consistent with the actual requiremnets of the heating surfaces.

The size of steam mains for heating apparatus, of course, depends on the pressure of steam to be used, the distance it is to be carried, the temperature of the exposure of the heating surfaces and their extent, etc. As it is not my intention here to speak of steam used expansively, I shall endeavor to give sizes only for the *direct return,* or gravity-circulation apparatus. The sizes of steam mains and other piping for heating apparatus of course depend on the pressure of the steam to be used, the distance it is to be carried, the temperature and the exposure of the heating apparatus and their extent; and this applies equally to all kinds of heating apparatus whether high pressure or low pressure. It is not my intention however to speak here of steam used expansively, but to give sizes, for the direct return of the water of the gravity circulating apparatus just described, leaving the sizes of mains for steam used expansively to a later chapter in the book.

A well-arranged gravity circulation should be made to work at *any pressure;* for with its heating surface properly proportioned it can be made to at least partly meet the exigencies of fall, winter, or spring weather, by simply carrying a pressure suitable to the occasion.

To have the water of condensation return directly into the boiler, under all conditions of pressure, the main pipes *must be large enough to maintain the pressure of the boiler to within half pound, in every part of the apparatus.* The water-line of the boiler should be not less than 4 feet from the bottom of the horizontal distributing mains at their lowest part, and that distance will only answer in short mains, such as those used in the generality of city business buildings and blocks. In large public buildings and others, having their boilers in out-houses, the difference between the boiler line and the mains should be all it is possible to make it.

A main steam pipe should not decrease in size according to the area of its branches, but very much slower, and should be rated by the heating surface and the distance steam is to be carried. Neither should the main at the boiler be equal to the aggregate size of all its branches—an expression very much in vogue in specifications for steam heating—or the mains may be unnecessarily large at the boiler.

Mains which have given the best results leave the boiler of sufficient size (calculated from practical results), and are reduced slowly, being proportionally larger the farther they have to carry steam, friction of the steam as it passes through them being an important element.

The area of the cross section of a 1-inch steam pipe is generally taken as unity in the rating of steam pipes for heating apparatus, and the area of a 1-inch pipe (in the main at the boiler) to each 100 square feet of heating surface is considered the limit of good ordinary practice. It has been the writer's ready method for years, and is deduced from the size of the mains and heating surfaces of some of the best heated buildings in the United States, and forms a safe approximation when closer methods are not at hand. The method of determining the size of steam mains for gravity apparatus by other methods will be treated in another chapter.

When the main steam-pipe leaves the boiler, it should, if possible, be carried high at once, and have the stop-valve at the highest part of the pipe, so that condensed water cannot lodge at either side of it when shut. This will prevent cracking at this part of the pipes when the valve is opened. If this arrangement cannot be carried out, and the valve has to be nippled on the dome of the boiler, or if there are several boilers, and they have to be made interchangeable with regard to their use, there should be a *relief* of large size in the main, just outside the valves.

It is well to mention here that a relief which leaves the steam-pipe must be brought into the return pipe in a position corresponding exactly to where it leaves the main; that is, when it comes from the outside of the main stop-valve, it should be taken to the outside of the main return valve. Otherwise, if an attempt is made to shut off, and both valves are closed, the water will "*back up*" and fill the apparatus. So, also, with all branches, risers or connections. If there is a valve in the steam part, *there must also* be one in the return parts corresponding thereto, and reliefs must leave the steam-pipe and enter the return on corresponding sides of the respective valves.

From the highest point the main steam-pipe should drop slowly, as it recedes from the boiler ($\frac{1}{2}$ inch to 10 feet being a fair pitch), that the course of the steam and the water may be in the same direction.

A main steam-pipe should not run very close to the wall up which the risers go. There should be room enough for a riser connection (2 or 3 feet), and when the mains are long, and the expansion great, the distance should be increased.

The T's in the main, for the riser connections, are better turned *up* than sidewise, as by nippling an elbow to them you can get any desired angle, and should the measurement for the main be a little incorrect it will assist in making connections. This arrangement also makes a good expansion joint, if the mains have much travel.

Where the pipe reduces in size, it is well to put a relief in the lower side of the reducing fitting, as the water that is pocketed there, by the large pipe pitching in the direction of the smaller one, may be the cause of cracking and noise in the pipe. Some steam fitters use an eccentric fitting in reducing which brings the bottom of the pipes on the same line and makes good work. See Fig. 8.

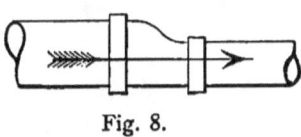

Fig. 8.

When it is necessary to have stop valves to the risers, the steam-fitter often places them in the riser connections, with a valve also in the riser relief. This arrangement requires three valves, and also stops the local circulation and equalization of pressure between the main steam pipe and return pipe when the valves are closed. Fig. 9 shows this method.

It is better to use only two valves, when it can be done, one to the steam and one to the return riser, and place them a few inches up the riser, above the riser connection, which brings them also above the steam-riser relief, saving a valve and lessening the chances for noise in the pipe. It allows the local circulation to go on between steam and return pipe when the valves are closed. This is shown in Fig. 10.

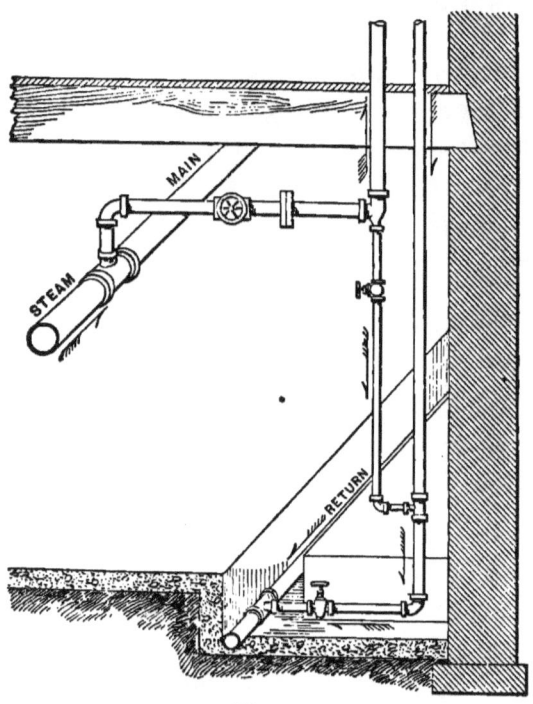

Fig. 9.

In System No. 2, where the returns are carried down separately, and collected together below the water-line, the return valve should be below all such connections, and the *steam-riser relief* should have a separate connection with the *main return*, and have no valve. This is shown in Fig. 11. Straightway valves are best for risers.

The extreme end of a steam main should be connected by a *relief* with the main return, being in fact, a continuation of the main down and into the return.

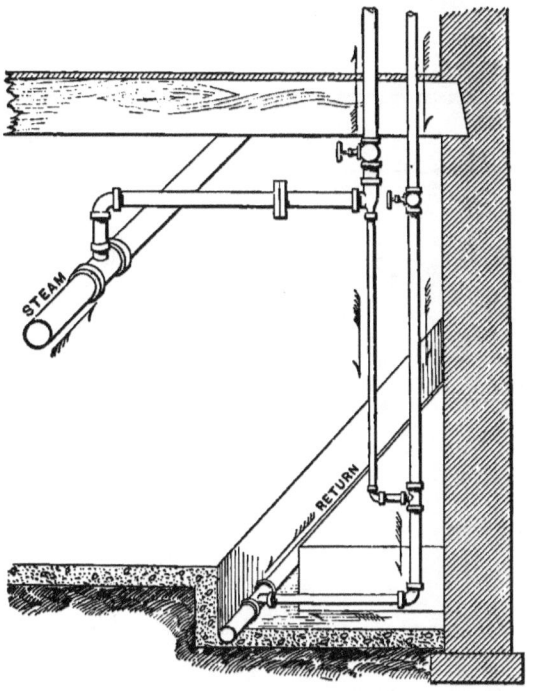

Fig. 10.

Stop-valves in main steam-pipes are either globe, angle, or straightaway. When a globe valve is used, it should be turned with its stem nearly horizontal, as shown in Fig. 12. The reason for this is obvious, when we consider that the water of condensation in any pipe runs along the bottom of it. When a globe valve is turned up, as in Fig. 13, the water in the pipe has to half fill it before it flows over the valve seat to pass along in the pipe. But, when the valve is on its side, it is different, for then *the side of the opening of the valve seat is as low as the bottom of the pipe.*

Neither should the stem of any valve be quite horizontal when it can be avoided. It should be raised

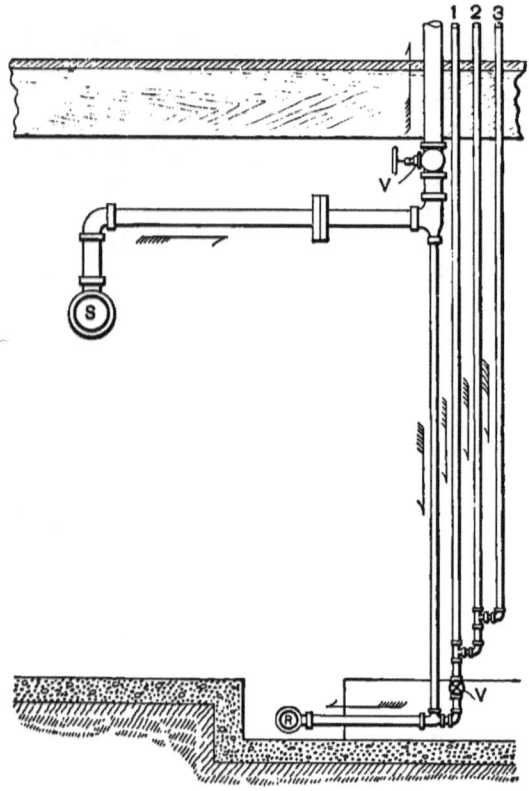

Fig. 11.

enough (10 degrees) to prevent water from collecting in the threads of the nut and stem, and being forced out by the pressure of the steam through the stuffing-box, which makes a constant dropping of water, that it is almost impossible to hold with ordinary packing. But with dry steam it can be held.

Globe or angle valves should be so turned in a heating apparatus that by simply closing the valve *to be*

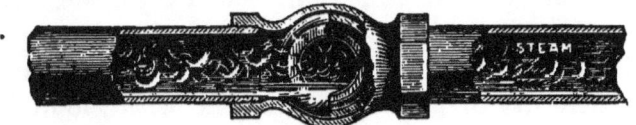

Fig. 12.

packed, and its *corresponding valve* in the return, or *vice versa*, and waiting for the steam to cool down, the stuffing-box or gland can be removed without the escape of steam. To do this it is necessary to have the *pressure side* of every pair of valves turned toward

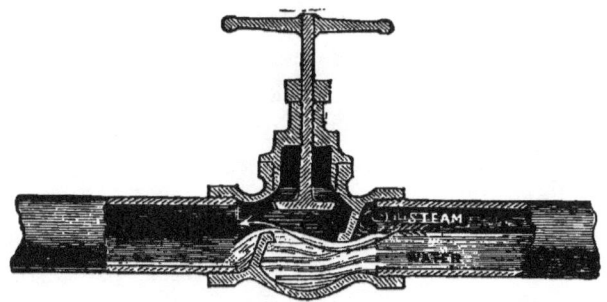

Fig. 13.

the boiler. By the pressure side of a valve is meant the under side of the disk.

Main Return Pipes.—In small apparatus (up to 3-inch steam-pipe) they are usually run on one or two sizes smaller than the corresponding steam-pipe.

In returns which are below the water-line, or are trapped to give them an artificial water-line, and consequently always full of water, smaller piping will generally do, but good practice has placed it at not less than one quarter of the area of the steam-pipe, for all conditions, for apparatus with larger than a 3-inch steam-pipe.

In apparatus with less than 3-inch pipe, the return is usually only one size smaller than the steam-pipe, that it may have a practical magnitude, and thus avoid the possibility of getting it stopped with the dirt or sediment carried to an elbow with the current of the water.

In dry returns—*i. e.*, which have no water-line—there are local steam currents, often going in contrary directions, the water gravitating toward the boiler, and the steam flowing to the heaters and alway in the direction of least resistance. This style of return is now much used in exhaust steam heating and in cases where there is no basement it cannot always be avoided even in gravity work. One-half the area of the steam-pipe has been found, in practice, to give good results in dry return pipes.

Check-valves are generally used in return pipes where they enter the boiler. Some steam-heaters leave them out on account of the back pressure they cause to the return water; but the practice is not to be recommended when two or more boilers are connected, as an inequality in draught, or the cleaning of a fire will make a small difference of pressure between boilers, causing the water to run from one boiler to another through the return pipes.

Check-valves of large area in the opening, with a small bearing on the seat, can be made that will not give more than one eighth of a pound back pressure.

or swinging check-valves can be used that practically cause no resistance to the flow of water.

The diameter of the return pipe is sometimes reduced where it enters the boiler, but it must only be done with caution and by one who is acquainted with the subject, as the resistance to the flow of the water increases much more rapidly than the decrease of diameter of the pipe.

Extra strong pipe and fittings should be used in all returns and feed-pipes, from where they are tapped into the boiler, to outside the brickwork; and when they are exposed to the action of the fire it is well to cover them with a "slip tube" made of a larger size, ordinary steam pipe.

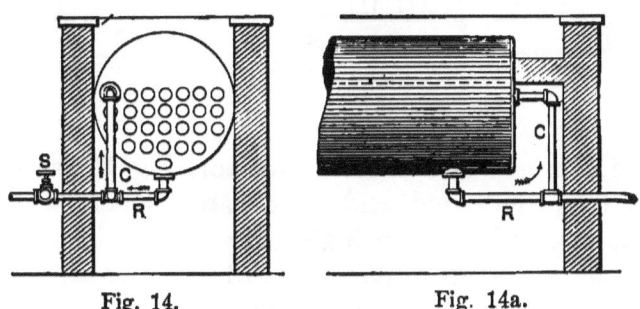

Fig. 14.   Fig. 14a.

A circulating pipe is sometimes used, which connects with a return pipe inside all valves, and again connects with the boiler head below the water line.

This is shown in Figs. 14 and 14 a. R is the regular return pipe, and S the stop valve, while C is the circulating pipe showing the direction of the water circulation when valve S is closed and the boiler used for power or some purpose other than heating. The circulation through pipe C prevents R from burning.

CHAPTER II.

RADIATORS AND HEATING SURFACES.

ALL radiators, box coils, flat coils, plate or pipe surfaces, arranged to warm the air of buildings, are *heating surfaces.*

The vertical wrought-iron tube radiator, and the cast-iron sectional type of vertical loop radiators, are now the accepted of first-class heaters, and nearly all manufacturers have their own peculiar style, with varying results as to efficiency. The steam-fitter or purchaser should use great caution in the selection of radiators.

Fig. 15.

The common return bend radiator, Fig. 15, is of old style, is not patented, and its construction is simple; a base of cast-iron, A, being simply a box, without diaphragms, with the upper side full of holes about 2¼ inches from centre to centre, tapped right-handed; a pipe, B, for every hole, 2 feet 6 inches or 3 feet long, threaded right and left handed, and half as many return bends, C, as there are pipes tapped left handed.

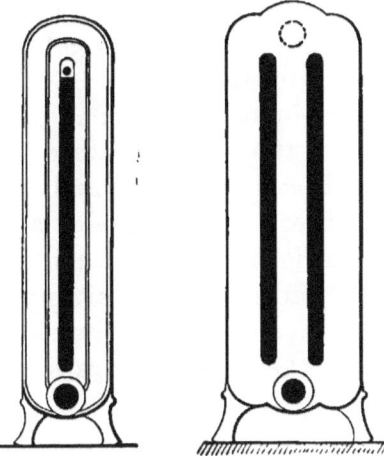

Fig. 16.   Fig. 16A.

The common manner of putting these heaters together is to catch the right-handed thread of two pipes one turn in the base, then apply the bend to the upper and left threads of the same two pipes, and screw them up simultaneously with a pair of flat-backed tongs on each pipe, while a second person holds the bend with a wrench made for the purpose.

The sectional cast-iron radiators are of many types, the principal feature of which is a cast-iron loop or section, two modifications of which are shown in Figs. 16 and 16A. These sections are joined together at the lower ends by screwed nipples or by compound tape nipples, and are usually reinforced by a long bolt running from end to end near the top, as shown in Fig. 17.

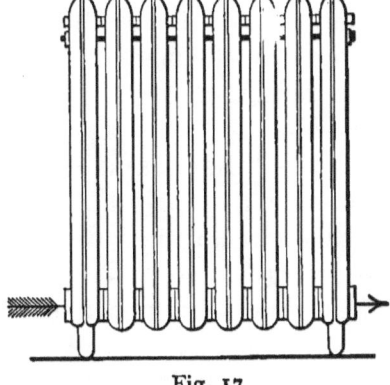

Fig. 17.

There are a great many designs of this cast-iron sectional radiator, and the principal difference that exists between them is in ornamentation alone, if we except the methods of joining the sections together. There are modifications also of the surface of the radiators, the object being to increase the condensation by the nature and form of the surface. Some are entirely plain surface, except ornamentation, and the vertical lines are straight; others have extended surfaces or partially extended surfaces, so that when the sections are put together they form vertical air pipes between the sections; others also have pin prejections to increase their surface area, the object always being to increase the efficiency.

Experiment has not always demonstrated that an increase of surface is an increase of efficiency, in that it sometimes shows the reverse, and the radiators that give the greatest efficiency per unit of surface are those of considerable distance between the loops, and have loops of a rather plain design. The effort of the manufacturer to-day is to increase the efficiency without increasing the floor space occupied by the radiator, and, if possible, not to increase its weight per unit of efficiency. This results in many forms, some radiators being long and narrow, others short and wide.

I will here explain the action of steam entering a radiator, as nearly all the patents on the so-called positive circulating radiators are to facilitate the expulsion of the air and the admission of steam.

The general impression among steam fitters is that when steam enters a radiator the air is forced up and confined in the tops of the pipes, which is the fact when the pipe is single, of small diameter, closed at the top, and without any of the usual means to get it down, although the air at the same density is about twice as heavy as the steam, and apparently should fall without assistance.

When two pipes are connected at the top with a bend, or when there is an inside circulating pipe, or a diaphragm of sheet-iron slipped into the heating pipe, the air immediately gives way and falls in the pipes nearest the inlet. Should there be no air-valve on the radiator, the air will be crowded to the return end of the radiator, and should the system be a gravity circulation, without an outlet to the atmosphere, it will remain in the radiator, impairing its efficiency and often deceiving the novice, as it in time heats somewhat by contact with the steam. When there is a thumb-cock or air-valve on the radiator, usually on the furthermost pipe from the inlet, the air can be withdrawn from the heater, when the result is quite different, the radiator becoming steam hot over its whole surface. In a radiator of good construction the action is direct, the pipes, loops or sections heat consecutively, excepting, perhaps, the one the air-valve is on, and a few near it, which sometimes heat ahead of their order, on account of the draught of the air-valve. Thus, when the steam enters a well-constructed radiator the air falls to the base and is driven out at the air valve.

In Plate I., at the front of the book, are shown some old ideas in regard to radiators, and the belief held about the necessity of making them positive in action. In No. 1 the pipe D is carried down nearly to the bottom of the base, the idea being to facilitate the circulation of steam and the expulsion of the air.

No. 2 shows a device (patented) for making a return-bend radiator positive. The pockets A A, filling with condensed water, makes a seal which at times prevents the flow of steam along the base and forces it in a continuous stream through the pipes (see arrows in cut).

Nos. 3 and 4 show cross-section of modifications of positive return-bend radiators. No. 3 can be used as a vertical radiator only, but No. 4 can be used in any position from perpendicular to horizontal, as seen at Nos. 5 and 6.

Single-tube radiators, welded or closed at the top with a cap, with an inside circulating device, were also much used; some of them compared favorably with the return-bend radiator, but were slower in heating.

No. 7 shows the first radiator put on the market, and one which has survived and is in use at the present time. A is the cast-iron base, B the welded tube, and C the septum of wrought-iron slipped inside the tube and projecting an inch into the base. This heater depends on the gravity of the air for a circulation.*

No. 8 shows another heater of this class which is positive in its action. A, cast-iron base; B, diaphragm cast in base; C, welded tube; D, inside tube, open top and bottom, and screwed into the diaphragm. The action of the steam can be seen by the arrows.

No. 9 shows a fire-bent tube radiator positive in its action.

---

* This was the original "Nason" radiator. An improved form of this radiator (Fig. 18, on next page) is now on the market, in which the bases of the double rows and all wider are perforated at the lower end of every tube, to provide for a more free circulation of the air.

23. Cast-iron radiators are of two kinds, *plane* and *extended* surfaces.

Plane surfaces, as the trade understands them, may be either flat, round, or corrugated, provided the coring, or inside surface of the iron, corresponds and follows the indentations of the outside, as in No. 10, Plate I., and in all wrought-iron heaters. Extended

Fig. 18.

surface is understood when the outside surface of the heater is finned, corrugated or serrated, with the inside straight, as in No. 11, Plate I.

For direct radiation, where the heater is placed in the room, there is little or nothing gained by having the surface of the heater extended, and a steam-fitter, in calculating the extent of his heating surfaces, should not take into consideration the whole outside surface of such a heater unless he gives it some value less than unity, the *unit* being a square foot average of plane surface.

For indirect heating (the coil being under the floor or in a flue) the result is different as compared with *shallow* plane surface coils, where the air cannot stay long enough in contact with them to get thoroughly warmed, but presses into the room without hindrance. In this case the extended surface gives a better result, though not because a square foot of the surface can transmit as much heat in the same time, but because it hinders the direct passage of the air, breaking it up and holding it longer in contact with the surface.

Fig. 19.

The cast-iron vertical loop or sectional radiator is a quick heater, the large size of the chambers facilitating the expulsion of the air.

Fig. 19 shows the "Bundy" cast-iron *loop* radiator. It is a cast-iron base with air apertures through its bottom, and the pipes are double tubes of diamond shaped section cast in one loop and joined to the base by a single thread.*

---

* The writer with his own hands made the experimental Bundy radiator in the shop of Mr. Chas. Gregg, in 62 and 64 Gold Street, New York. It was the first cast-iron loop radiator,

Fig. 20 is the "Reed" cast-iron loop radiator. The bases are apertured for the passage of the air, and

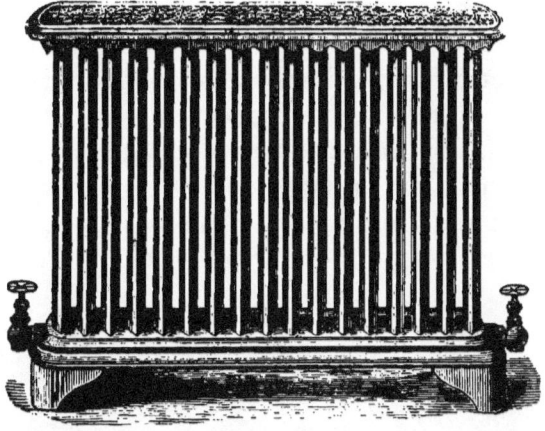

Fig. 20.

the loops are U-shaped tubes fastened at each extremity to the base by the assistance of copper ferrules, into which the loop is forced by pressure. The bends

Fig. 21.

of the loops, at the top, interlock and are confined by a rod to prevent disturbance in handling.

Fig. 21 shows the old box coil, the primitive form of an indirect radiator.

Fig. 22 showed the first extended surface radiator for indirect work. It is known as the "Gold" pin radiator, each section, for steam use, containing nominally 10 square feet of heating surface. I show these two indirect radiators here as the box coil is the representative of the plain surface indirect radiator, and the Gold pin the representative of the extended surface radiator. In a later chapter of the book the question of radiators will be gone into more fully.

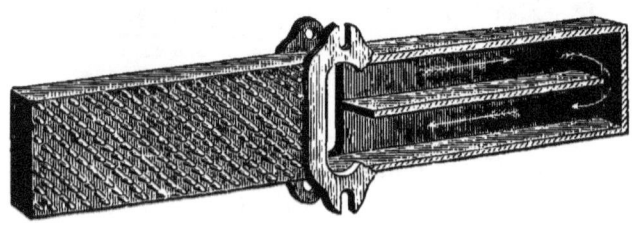

Fig. 22.

Fig. 21A shows a coil of secondary surface known as "Gold's compound coil surface," and which was used for indirect heating, either as a steam coil or as a hot-water coil. It is also made into direct heaters in the form of one-inch vertical pipe radiators covered with the *secondary* surface, and inclosed within a sheet-iron case, with a register in the top to control the heat.

Fig. 21B shows an inch pipe covered with this surface, which is No. 14 square wire in the helical form, one pound being wrapped on each lineal foot of pipe. It increases the efficiency of the one-inch horizontal pipe in condensing power full *three* times when made into indirect coils and properly boxed, and

Fig. 21A.

in vertical radiators of moderate height the efficiency is about double what it would be for plain pipe.

Sheet-iron radiators are used in very low-pres-

Fig. 21B.

sure heating, the commonest form of which is the flat Russia-iron heater, seamed at the edges and studded or stayed in the middle, with a space of about $\frac{3}{8}$ of an inch between the sides. They are used in a one-pipe apparrtus and are occasionally seen in old buildings.

Coils are always made of wrought iron steam-pipe and fittings, and though not considered very ornamental, are first-class heaters, and give a high efficiency of condensation per unit of surface.

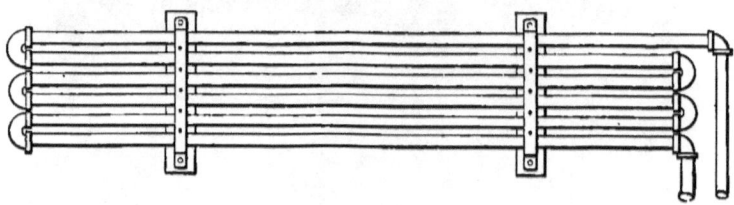

Fig. 23.

Fig. 23 show a *flat coil*, which is a continuous pipe, connected with return bends at the ends, and strapped with flat iron, or supported on ornamental brackets. It is a very positive heater.

Fig. 24 shows a miter or wall coil. It is composed

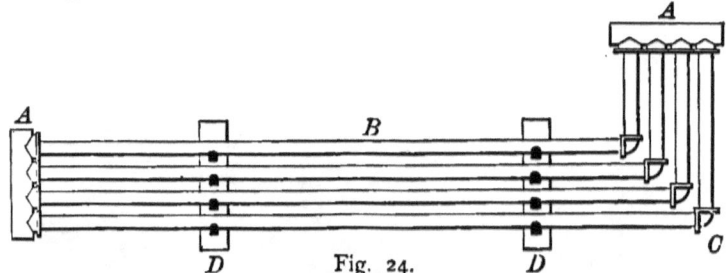

Fig. 24.

of headers or manifold, A A ; steam-pipes, B ; elbows, C ; and hook plates or rosettes, D.

There are many modifications of this coil, and they are usually long 20 to 30 feet. One indispensable point in the making of this coil is it must *turn a corner* of the room, or miter up on the wall as shown. The pieces from the elbows to the upper header are called *spring pieces;* they are screwed in right and left, and are the last of the coil to be put together.

If a coil is put together, straight between two headers, as seen at Fig. 25, it will be like Fig. 26 when heated, and cannot be kept tight for a single day; the

Fig. 25.

expansion of the first pipe to heat, being a powerful purchase to force the headers asunder, and when it cannot do so it will spring the long pipes out of the hook plates or force the pipe to strip its own thread. The modification of the wall coil, shown in Fig. 24 is that used when there is no chance to turn a square corner of the room. A miter is then made in the

Fig. 26.

manner shown. When advantage can be taken of the corner of the room, however, it is better to turn a right angle on a horizontal plane, and in this way let the expansion of pipes from both directions be forced into the corner. The supports at the corner should be placed so that the hook plates of the short side of the coil will be as far as possible from the corner, and the hook plates and the long side very near the corner. This will allow for the movement of the long pipe of the coil, as it expands to move towards the corner by slipping through its own hook plate, allowing all the pipes to compensate and equalize on their own supports without undue strain, and at the same time providing a support very near the corner of the coil, which, of course, is necessary to maintain the pipes in their proper alignment; a corner being more apt to hang down than a straight run of pipe. The movement of the short part of the coil being small, it is easily compensated, so that if the support near the long pipe is but one foot from the corner, there would be no trouble in taking care of the expansion of five or six feet of pipe should the spring piece (short part of the coil) have that length.

There are many other modifications of the wall coil, the particulars of which it is not necessary to enter into here.

CHAPTER III.

CLASSES OF RADIATION.

HEATING surfaces are divided into three classes: 1st, direct radiation; 2d, indirect radiation; and 3d, direct-indirect radiation.

Direct radiating surfaces embrace all heaters placed within a room or building to warm the air, and are not directly connected with a system of ventilation.

The best place within a room to place a single radiator, is where the moist air is cooled—namely, before or under the windows, or on the outside walls. When the heater is a vertical tube radiator, or a short coil, which can occupy only the space of one window, and when, as often occurs in corner rooms, there are three windows, the riser should be so placed as to bring the line of radiators in front of, and under the windows where they will do the most good—as the middle window. Or it is better still, when a small extra cost is not considered, to use two heaters, and place one in front of each of the extreme windows.

When the room is large and has many windows, the heating surface, when composed of radiators, should be divided into as many parts as possible, and, perhaps, as there are windows; or should the owners or occupants object to so many windows being partly obstructed, divide into half as many parts, and distribute accordingly.

In schools or factories, or buildings with many windows, where children or persons cannot change their positions, but have to remain seated for several hours at a time, care must be taken that the heating surface is very evenly distributed. A coil run the whole length of the outside wall is best, but if any kind of short heaters are used, every window, or at least every second window, should have its share of the heating surface. Should a single window be left unprovided for, it will be found by experiment that a cold current of air will fall down in front of such window and flow along the floor in the direction of the nearest heaters, and probably cause cold feet to any who are in the line of its passage.

The natural currents in a room with the outside atmosphere the coldest, are always down at the windows and outside walls, and up at the center of the room or rear walls. This downward and cold current, should be met by the heated and upward current from the radiator, and reversed and broken up, as much as possible.

Indirect radiation embraces all heating surfaces placed outside the rooms to be heated, and can only be used in connection with some system of ventilation.

There are two distinct modifications of indirect radiation. One, where all the heating surface is placed in a chamber, and the warmed air distributed through air ducts, and usually impelled by a fan in the inlet or cold air duct. The other, where the heating surface is divided into many parts, and placed near the *lower ends* of vertical flues, leading to the rooms to be heated.

The first of this class—namely, *chamber-heat*—has not proved satisfactory unless the air is impelled with a fan, as it has been found that in windy weather it is almost impossible to force air to the side of a building against which the wind blows, with natural currents alone. With the proper fan, however, this method of heating has recently become the one almost entirely used of late years for hospitals and schools and many public buildings. There are many modifications of it, and the subject will be treated in another chapter of this book, showing its many modifications and general application to buildings. The second of this class does better than the first without a fan, as it admits of taking advantage of the force of the wind, to aid in bringing the warmed air into the rooms, and does very well for private houses. The second and well known method of indirect radiation is the ordinary one of having a radiator at the base of a vertical flue, the cold air being taken directly from out of doors and delivered in the room through the warm air pipe and register, and this method has always done well for private residences and where the quantity of air that carries heat into the room is ample for its ventilation, as in medium size rooms or offices where a very few persons only are occupied or engaged.

The indirect heater is usually boxed, either in wood lined with tin, or in sheet metal. The former is best when the cellar is to be kept cool, as there is a greater loss by radiation and conduction through metal cases; otherwise metal is best, as it will not crack, and when put together with small bolts can be removed to make repairs, without damage.

The vertical air ducts are usually rectangular tin or galvanized iron flues built into the wall when the building is going up. Sometimes, however, they are only plastered in the rough, and sometimes lined with tile, but the smooth metal lining with close joints gives much the best result in all cases, and in the case of outside walls it is about absolutely necessary that the flue should be lined, as the passage of air and consequent loss of heat through a porous wall is very great; moisture also playing a considerable part in destroying the efficiency of the flue in damp weather.

The cross section of an air duct should be as large as it is possible to obtain in the construction of a building, as large bodies of air moved slowly give very much better results than when the velocity has to make up for the volume.

A $12'' \times 12''$ flue in a wall will deliver between 10,000 and 15,000 cubic feet of air in an hour to a room on the second floor of an ordinary house, if it has easy bends and is not too much obstructed by the radiator or register, and about one-half that amount will be delivered under similar conditions to the first floor through a short flue or floor register. A good common rule is to make the first floor flue twice the area of the second floor flue.

There should be a separate vertical air duct for every outlet or register. In branched vertical air ducts uniformity of delivery is almost impossible.

The heated air from one heater may be taken to two or more vertical air ducts when they start directly over it or very near it; but one should not be taken from the top, and the other from the side, or the latter will be unsatisfactory unless the room to which the flue runs is exhausted; *i. e.*, the cold or vitiated air of the room is drawn out by a heated flue or fan.

Inlet or cold air ducts are best when there is one for every coil or heater, in what may be called the private house method; and its mouth, or outer end, should, if possible, face the same way as the room to be heated. By this means, when the wind blows against that side of the house, the pressure is into the cold air duct, and materially assists the rarefied column of air in the vertical duct to force its way into the room.

Often the steam-heater uses only one large branched cold air duct; but this system may give trouble unless the trunk and branches are carefully proportioned.

The steam-heater should not undertake a job of indirect heating unless the building has been arranged especially for it, with some efficient system of flues, sufficient to change the entire air at least once in an hour.

Frequently designers make no provision for drawing out the cold or depreciated air. Such rooms cannot be warmed by indirect heating at all. But when there is a chimney or an unwarmed outlet or foul air flue, the heated column of air in the vertical hot air flue is generally sufficient to force its way through and into the room in quantities sufficient to warm it, even though it may not ventilate it to any considerable extent.

A cheap good way to make outlet or foul air flues, draw or to exhaust, is to connect them all to one large annular flue, around the boiler chimney flue.

Warmed fresh air flues should be near the outside walls, but not in them, if possible, and should discharge near or across the windows; and foul air flues should be in the inner walls, and have an opening near the floor and ceiling, with register valves, to allow the occupant to use either, or both, as he thinks proper. Usually the lower vent is without means of closing it.

The velocity of the air in heating flues with only a natural draught, rarely reaches 8 feet per second, no matter what the condition; and 2 feet, 4 feet, and 5 feet respectively, are fair averages of velocities for first, second, and third floors of a house. Of course when the doors of the rooms are shut and there is a systematic arrangement of vent flues to the top of the house, the velocity of the draught for the short flues is considerably increased, as under such conditions the combined effort of both heat and vent flue tends to accelerate the current of air in the warm air flue.

Direct-indirect radiation embraces all heating surfaces placed within, or partly within, the room to be warmed, in direct connection with some system of ventilation.

Heaters of this class are usually placed on the outside walls or under windows, following the same general rules as for direct radiation, excepting the clusters are generally deeper, so as to prevent the cold air from rushing through without being warmed.

Fig. 27 is a favorite modification of this style of heating. It is a section of a room, showing the apparatus and the supposed action of the currents of air. $A A$, outside wall; B, partition wall; C, radiator; D, inlet flue; E, damper or valve; F, ventilating flue or foul air outlet; *G*, fresh air mixing with the air of the room; *H*, air of the room passing along the

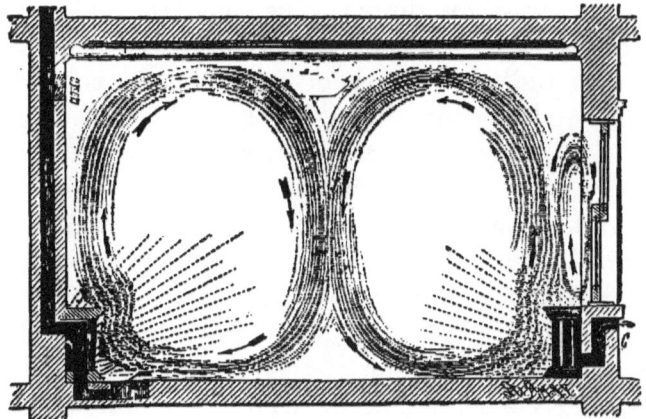

Fig. 27.

floor to the heater; *I*, a percentage of the air in the room passing off by the ventilator.

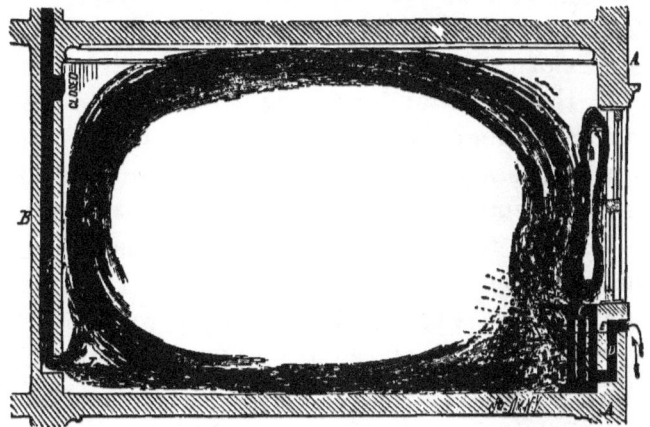

Fig. 28.

Fig. 28 is a modification of direct-indirect radiation not often met with, but where with good success some of the *local heat* is employed to exhaust or draw out the vitiated air of the room and draw in the fresh air. The arrows show the supposed general action of the air currents within such a room. *A* is a section of a radiator built with a sheet-iron flue, *B*, between the tubes, and passing through a hole, cored in the base of the radiator which connects with the register in the floor, and a foul air flue in the wall. This system might be used to advantage in schools where a fan cannot be employed.

Some of the radiant heat, etc., *A*, warms the sheet-iron flue, *B*, which in turn warms the air within it, causing an acceleration of the current in the foul air flue, and consequently drawing an equal amount of fresh air in at the opening, *C*.

In estimating heating surfaces, for direct-indirect heating, it is well to use about *once and a half* as much as would be used for direct radiation alone.

There is this further distinction between the three systems of radiation: Direct radiation warms only the air of the room and maintains the heat. Indirect heating warms only the air that passes in, and cannot warm the same air twice, and consequently has to raise the temperature of all the air that passes, from the outside temperature to that necessary to maintain the temperature of the room. Direct-indirect radiation warms part of the air of the room over again, and warms all the air admitted for ventilation, which latter can be varied to the point of shutting it all off and depending on direct radiation alone.

With ordinary indirect radiation (no fan), the heating apparatus being steam, a building *may* be sufficiently ventilated; but it frequently happens in large rooms with very high ceilings, or large auditorums, as churches, schools, theaters, or assembly rooms of any kind, that they are not always satisfactorily heated, as it is difficult to warm them by indirect radiation alone, unless there are many registers placed before the windows, or the apparatus is supplemented by direct radiators, placed where there will be strong local currents.

Heated air from a few large registers in a very large room goes directly to the ceiling, and fills the room from above, expelling the same amount of air through the ventilators. If the building had no windows, this would probably answer; but as buildings have windows and outer walls, which cool the air rapidly, there will be a falling of air in front of the windows, etc, which has not been pressed down by the warm air above, but has fallen of its own gravity, by losing its heat from contact with the *cooling surfaces* of the building. These downward currents, having nothing to neutralize them, pass over the heads and shoulders of the sitters and go cold along the floor on their way to the ventilator, or to an ascending current of warm air, caused by the heat given off from the bodies and lungs of the audience or any other cause.

This is why people in churches and theaters suffer from cold backs and feet, and sometimes have a cold current on their heads, which makes them certain "the window is open a little;" though a thermometer near by marks 70 degrees, as the thermometer is probably not in the cold current and will not take note of draughts anyway.

If a building must be heated entirely by indirect radiation, it is well to use as many heat registers as possible, and place them in front of the windows, or where a cold current is likely to come down.

Usually in office rooms, and ordinary rooms in residences, *one* register in the coldest part of the room can be made to answer; but if the room is large, with many windows, more should be used.

Figs. 29 and 30, perspective elevation and section respectively, of one of the indirect radiators in the Cambridge Hospital, and show the arrangement of the air-inlet pipes *A*, mixing valve *D*, hot-air pipe *H*, and register-box *R*, within the wards, as well as a section

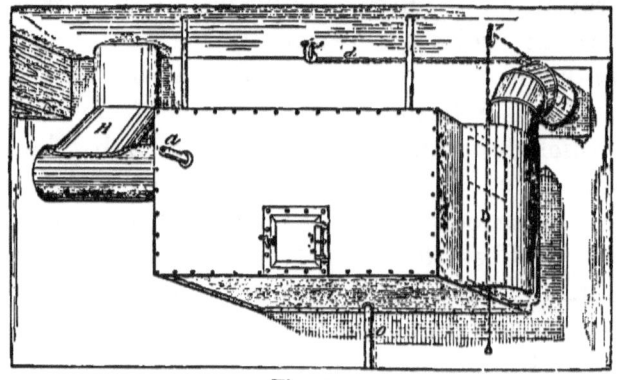

Fig. 29.

of the vent-ducts, *V A*, with the vent-outlet under each bed (*V*).

The coil casings or air boxes are made of No. 22 galvanized iron, with flanged corners, and the steam-radiator is suspended midway in the case, as seen in Fig. 30. The indirect heaters are "pin" sections, centre connection, eight sections being used to each hot-air box. The cold air enters through the 10-inch round pipe *A*, Fig. 29, and shown separated on cellar plan by the arrows, the mouth of which is protected by a register-face and frame. As the air enters through *A* it can be made to pass either under and through or above the heating-surfaces of the radiator by means of a sliding damper, *D*, or the air-current may be divided by placing the damper in a nearly central position, allowing some of the air to pass each

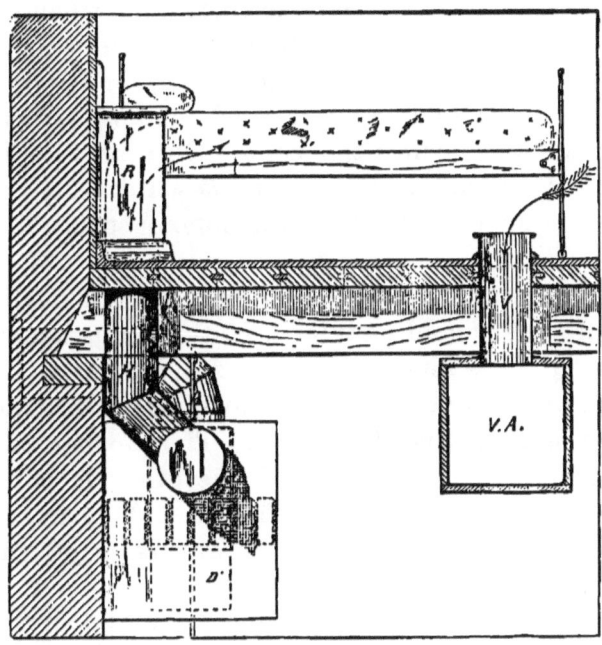

Fig 30.

way, thereby regulating its temperature without reducing its volume. The upper end of this damper is connected by a chain with a pull-and-stop mechanism within the ward, so that the attendant can regulate the heat of the air without leaving the room. The apparatus is called a "mixing-valve." This modification was selected, as it embodied the ideas of Dr.

Wyman, a well-known writer on ventilation; but more modern examples will be given elsewhere.

It will be seen that the dampers in the cold air inlets are not automatically regulated. They are sometimes so regulated to prevent the freezing of the coils; but when the steam and return pipes are sufficiently large coils, are seldom frozen, for when steam is up they cannot freeze, and when steam is not up, there is no water in the coils to freeze, for it has subsided to the water line level, which should always be a safe distance below the coils. Only an apparatus with small diameter *pipes and parts* will freeze, unless the coil is too close to the water line, or partly below it.

Indirect coils, if they have valves, should never be shut off in very cold weather. If the room is not to be heated, close the registers and inlet ducts. The closing, or partly closing of a valve, may freeze a coil, by interrupting the circulation. The closing of one valve and the leaving open of the other is sure to freeze a coil if exposed to sufficient cold, as in either case it will fill with water. *This applies to all radiators.*

Fig. 31 shows a modern switch valve and arrangement of an indirect coil now usually applied to school rooms. The radiator is generally a large one, as it is usual and necessary to admit about one hundred thousand cubic feet of air per hour to a room that will contain fifty pupils.

The radiator, therefore, has to have a large air area. In other words, it is not so necessary to have it deep as it is to have it cover a large surface, in order that abundant air can pass through it without material

STEAM HEATING FOR BUILDINGS.

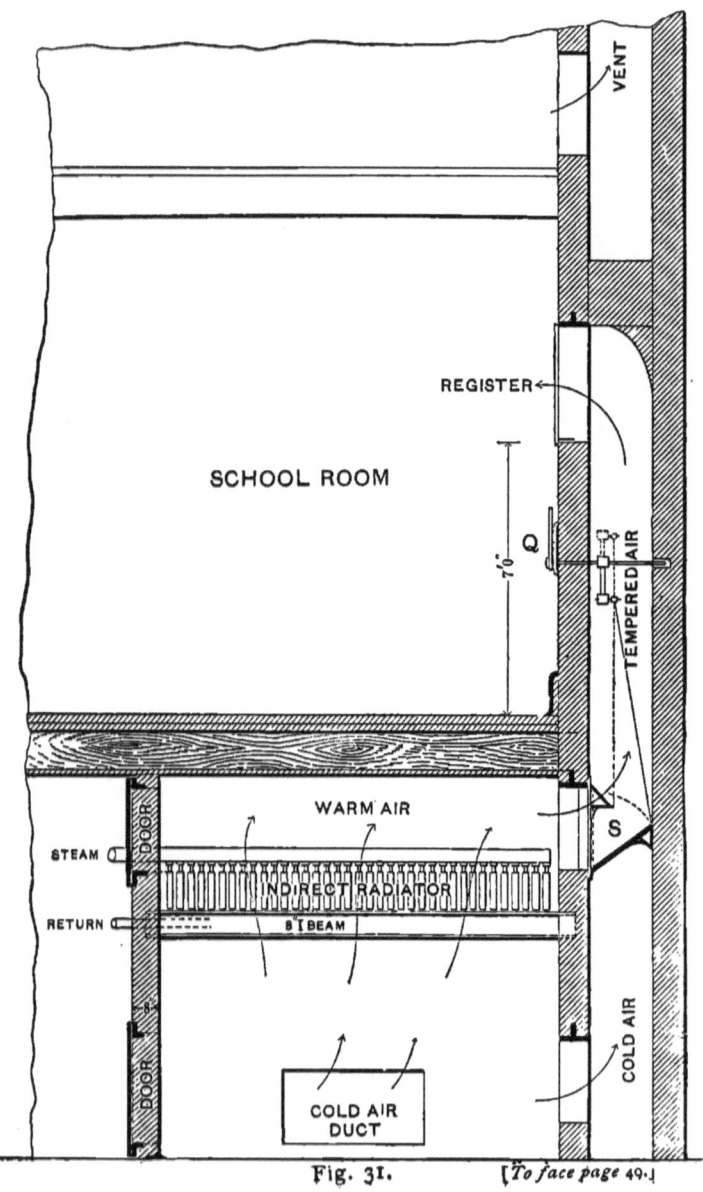

Fig. 31. [*To face page* 49.]

resistance. The flues, in like manner, have to be of ample area, $3\frac{1}{2}'$ to $4\frac{1}{2}'$ of cross-section generally being necessary for the average school room. The indirect radiators are built on I beams and enclosed in brick walls, though sometimes galvanized iron is used. Cold air is admitted at the bottom of the chamber, where it may either pass upward through the radiator and through the switch "S" into the hot-air flue and thence to the register in the school room, or it may pass through the lower opening into the same flue when the switch valve "S" is closed. At "Q" on the wall in the school room is a quadrant operating a rod which works a lever that connects directly with the switch valve "S" by a chain. When the switch valve is in the position shown by the full lines, entirely warm air is passing to the school room. When the switch valve is in the position shown by the dotted lines, cold air only can pass into the school room, but when it is put in any intermediate position, cold air from below the radiator and warm air from above it meet and mix in the flue, and pass as tempered or "mixed" air into the school room.

This is the usual method now of securing variations of temperature in the school rooms, without interfering with the quantity of air to be admitted in a given time. Details of construction are shown by the drawing.

CHAPTER IV.

HEATING SURFACES OF BOILERS.

The direct heating surface of a boiler (the fire box or crown sheet), has a value several times greater than the indirect surface (flue or tube surface), but the shape of the furnace, its size, and the angle of the heating surface, as well as the length, size, and position of the flues, give a greater or less value to the indirect surface; these values, of course, are only comparative.

In constructing boilers for heating apparatus, an effort should be made to have the greatest amount of direct surface, with a minimum of indirect surface; for it is desirable to have slow combustion, with thick fires, and thus reduce the attendance to a minimum.

When furnaces are comparatively small, with a high rate of combustion, flue surfaces may be lengthened with beneficial results; but in a private house, with a self-feeding boiler (base burner) or one which has a deep furnace, constructed to put in six to eight, or even twelve hours' coal, and keep steam uninterruptedly for that time, a great part of the heating surface should be in the fire-box; the heat from the gases being comparatively low tempered, and the amount passed in a given time small.

It would be well to say that most writers on boilers

put too high a value on what is termed *direct heating surface*, in contradistinction to *indirect* or flue surface. Not that a square foot of surface in a fire-box of ordinary construction has not $2\frac{1}{2}$ to 4 times the value of the same area of average tube surface, but they convey the idea, that by increasing surface near or in the fire-box, and decreasing the tube surface near or in the direction of the chimney in a threefold proportion to the increase in the fire-box, they can evaporate as much water with the decreased surfaces. Below certain sizes and proportions (which have already been attained in boilers of ordinary good construction), this may be so, but when a fire-box or furnace is large enough for proper combustion, the surface of it then receiving all the radiant heat *there is*, and by increasing the surface directly exposed to the action of the fire (beyond the required chamber for combustion), it will be necessary to have the surface of the fire-box as a whole, more remote from the fire; as radiant heat from any source has its effect decreased, *directly as the surface which absorbs it.*

From a central point of heat the rays diverge on all sides, and the intensity diminishes *inversely as the square of the distance,* which will be found to be *directly as the surfaces of different sized spheres, which might surround it.* The value of the heating surface (for radiant heat) decreases for each unit of distance, in a geometrical progression; in other words, twice the distance, one-quarter the heat. The above can be likened to the fire in an upright boiler.

In horizontal boilers, or boilers with long fire-boxes, or ones fired within horizontal cylindrical furnaces, the fire can be likened to a long column of heat, from which the rays go off parallel to each other in the line of its length, but diverge in a line of its cross section; which will give a progression whose ratio is 2 as the decreased value of the surface for each distance it is removed from the fire; in other words, twice the distance, half the heat. But in any case, the assertion that the intensity of radiant heat decreases directly as the surface which absorbs it will hold good for any shape of fire or any shape of furnace, and that hanging tubes, projections, or corrugations in a fire-box receive nothing from the radiant heat that would not be received by the plain surface; so, although a person may take 4 foot of tube surface away, and add one foot to the fire-box without perceiving they lost anything, yet they cannot, in a boiler that is already $\frac{1}{5}$ furnace, and $\frac{4}{5}$ flue, whose gases of combustion escape at a sufficiently low temperature, take away all the flues, or a large percentage of them, and by adding $\frac{1}{4}$ of their surface to the fire-box, makes as much steam.

All that can be gained by crowding the fire-box with surfaces, hanging or otherwise (which must not interfere with combustion), is, to reduce the bulk of the boiler; the surfaces will be the same still, for the same work. It is therefore poor economy to reduce the size, when nothing else is gained, and make surfaces which will fill up on the inside with sediment, choke up in the tubes, or between them with soot and ash, and wear out in one third of the ordinary time.

It is an incontrovertible fact that boilers with very small parts require more surface for the same work done than with large and plain parts, because of the impossibility to thoroughly clean them and the rapidity with which they choke, the nearness of the tubes allowing the dirt to *bridge* between them.

A maximum of fire-box with a minimum of flues is, however, proper, and should be the rule in house heating, where there is generally plenty of room in the cellar.

If the surface of the fire-box be increased by projections or corrugations, for the purpose of an increase of surface in contact with the highly heated gases of the furnace, the folds should be large and in vertical rows, so nothing can find a lodgment on them.

The boilers which have given the best evaporative results, as well as the least trouble, and lasted the longest, have been the simplest, and the evaporative results of a boiler depend more on the care with which they are kept clean, and the unimpeded circulation of the water within them, than on any *peculiar disposition* of the heating surface.

Large boilers, compared to the work, are most economical, but the limit is hard to fix. There are so many conditions to be taken into consideration, as well as styles of boilers, and as it is really the size of the grate and the velocity of the draft, compared to the work to be done (after the boiler is large enough to make sufficient steam), which regulate the economy hence a sufficiency of boiler with the proper *grate surface* to burn the fuel accomplishes the most satisfactory results.

A boiler that may do very well for the first year may not give satisfaction the second year. Such will be the case with boilers barely sufficient for the work, which, while they are clean and the person in charge of them has a pride in doing well, will pass muster;

but the second year, when the novelty has passed off, it will be quite different; then complaints will be heard, and one investigating steam apparatus with a view to putting it in his house, will be apt to reject it should he inquire no farther.

In proportioning the size of boilers for heating apparatus all calculations must be based on the supposition that the boiler will be neglected to a certain extent, and that there are parts of the best boilers which cannot be properly cleaned, and that all boilers deteriorate in transmissive power (the gravity return least of all, as the return water is *pure*) more rapidly at first, until a point is reached where external deposits fall off, after which the impairment is slow, and caused only by slight deposits on the inside, chiefly oxides, which have a high transmissive power themselves.

Can a boiler, it may be asked, be robbed of its heat by the gases of combustion, by retaining them too long in contact in passing through long flues? Not if they are internal tubes or flues; but there is a *point* beyond which there is no gain—namely, where the temperature of the gas and the steam becomes the same. Up to that point the gases of combustion, being the hotter, impart heat to the flue, but beyond it neither the flue can impart heat to the gas nor the gas to the flue, as they are of the same temperature. Boilers, when they are new, should have some such point, which simply moves nearer the chimney as they become old and dirty.

The rate of combustion will also give this point a variable position, for the time being.

Some engineers think it preferable to let the gases

of combustion escape at a higher temperature than the steam. In that case the point can be assumed to represent any *constant* difference of temperature of the gas above the steam.

Reverberatory, or drop flues, in upright boilers, save much heat. A cause of loss of heat, in upright boilers (and possibly in many other boilers), which have a great many tubes, many more than the aggregate area of the chimney, is that the heated gases find the tubes directly over the fire and pass out rapidly at a high heat, of their own gravity, leaving the gas in the outer rings of tubes inert, as may be seen in almost any upright boiler, where the tubes of the outer circles are generally found clogged with dirt; the velocity of the draft in the middle tubes keeping them comparatively clean. But when there is a row of drop tubes, as shown in Fig. 37, or a flue built around the outside of the shell of the boiler with brickwork, with the chimney flue leading from the bottom, as shown in Fig. 36, the gases are then *drawn out* or "exhausted" by the heat in the chimney; and the gases around the upper part of the boiler become uniform in temperature, and stratify, the lowest being drawn off first, and the others following according to their temperature.

When combustion is good, and the gases as they leave the boiler and enter the chimney flue have not too high a temperature, *the water within such a boiler has absorbed all the available heat;* hence, to increase the surface of such a boiler, will not do much good, unless the grate surface is also increased; since *all* the heat evolved has been absorbed.

NOTE.—Figs. 36 and 37 are in Chapter VI.

Will the quantity of water within a boiler effect evaporation?

Many steam heaters and others use boilers composed of very small parts, so as to have the greatest surface with the least water, with a view to evaporate more water in a given time, and cite the time *between starting the fire* and the time *steam is up* as a proof of it. This is a mistake! The reason why steam is gotten up quicker, is because there is less water to heat to 212° before steam begins to make, but beyond that, the result, with regard to steam making is the same, for the same surface, other things being equal.

What is gained in *first* time, with sensitive boilers, is more than compensated for, in house heating, by having boilers which contain a large quantity of water, that hold their steam when a new fire is put in. Boilers which contain small quantities of water are rapidly chilled, as well as rapidly heated, and must be fired often, and regularly.

Fire engine boilers require to be sensitive, and when much power with small weight is a desideratum they are all right, but they are not fit for house warming, nor are very sensitive boilers of any description.

CHAPTER V.

BOILERS FOR HOUSE HEATING.

Boilers for heating apparatus should have very few parts, and be as simple as it is possible to make them, every part of them being constructed with a view to permanency, and parts that wear out more rapidly, such as grates, should be so arranged that they can be renewed by the most inexperienced person.

Requirements for house heating boilers are:

1st. They should contain a quantity of water above the safe line sufficiently large to fill the pipes and radiators with steam, to any required pressure, *without lowering the water enough in the boiler to require an addition* when steam is up; for should the steam go down suddenly, there will be too much water in the boiler. This occurs in boilers made with very small parts or pipes, which have a small capacity. Should such a boiler have an automatic water feeder, set for the *true* water line, it will fill up, but cannot discharge again when the steam goes down; while if it has *no* feeder, there is danger of spoiling the boiler, as too great a proportion of the water is in the pipes *in the form of steam.**

* For the quantity of water necessary to fill the pipes with steam at any pressure, at a maximum density, see Table 13

2d. The fire box is better made of iron, with a water space around it, as in upright or locomotive boilers, to prevent clinkering on the sides and the necessity of repairs to brickwork; which are *unavoidable* in brick furnaces.

3d. The fire box should be deep, below the fire door; to admit of a thick fire, to last all night, and thus keep up steam for a long period.

4th. The fire-box should be spacious, for the sake of good combustion.

5th. The flues and tubes should be large, and in a vertical position, so they will not foul easily, and that any deposit may fall to the bottom or into the fire.

6th. The heating surface should be great in its diameter instead of in the direction of the chimney.

7th. They should, if possible, be constructed of such shape and design that they will require no sweeping, or cleaning, other than removing the ashes, but when it is unavoidable, every facility should be made for easy access to such parts, as they are often operated by inexperienced persons (house servants), who naturally find fault with anything that gives them trouble.

8th. The fire-grate must be easy to clean and so designed that it will not crack or break when heated. Grates of the shaking or rocking pattern only should be used, and they should not be too fine. (See article on grates).

9th. The grate and ash-door must be so constructed that a new grate can be put in quickly by any one.

10th. There should be no tight dampers in the chimney flue, and when the flue goes out near the bottom (drop flue), they may be dispensed with altogether; but the fire and draft-doors should be made to close air-tight, so as to be capable of entirely damping the fire. This will prevent the possibility of coal gas escaping into the house. The damping of a fire by shutting off its supply of air, is the proper way for house work, as the draft of the chimney being unimpaired, it draws all the harder on any crack or crevice in the brickwork, causing an inward current, which entirely precludes the escape of gas into the house.

11th. The perpendicular height of the boiler should not be too great for the cellar, or the water line will be too near the main pipes or radiators.

12th. A boiler should be so enclosed in brickwork or asbestos or magnesia coverings as not to perceptibly raise the temperature of the cellar in which it is. This also makes the whole outside of the boiler heating surface if required, by having either an upward or downward flue.*

When upright boilers are constructed with drop tubes, as shown at a', Fig. 37, or with drop flues, as shown in Fig. 36, it is generally necessary to use a direct smoke pipe as well as a bottom pipe, as shown, in which case an upper damper is required and possibly it is better to have a lower damper also. The two dampers should be connected at right angles to each other by a rod, as shown at i, Fig. 37, which prevents the possibility of having both dampers closed together.

In upright boilers for house heating, the proportion of fire-box to the flue surface admits of almost any modification, as the boiler can be made of large diameter, with short tubes and high fire-box drawn in

---

* Asbestos-lined jackets of iron, or other suitable jackets of refractory materials, may take the place of the brickwork.

at the bottom with dead plates, for the desired size of grate, or drawn in as shown in Fig. 35.

Horizontal multi-tubular boilers admit of very little modification; a large diameter, with short shell and large tubes being best for slow combustion, with a considerable distance between the grate and boiler, and no bridge-wall higher than is sufficient to keep the fire on the grate.

A chamber behind the bridge-wall is not of any particular service, when the bridge-wall is low; and making a *contracted* throat, at the bridge-wall, or behind it, to make the heat "hug" the boiler, is a mistake. What is wanted in the furnace, and under the whole length of the boiler, is *space* sufficient for complete combustion. Below a certain size of cross section combustion is interfered with, and the oxygen which passes through the fire will not combine with the carbon unconsumed at the grate, but with ample space this ignition will be continuous until complete with a sufficiency of oxygen, where the temperature is not below (800°) eight hundred degrees Fahr.

For a high rate of combustion the boiler may be longer, with tubes of small diameter and with great space under the boiler.

A contracted passage, or one having only the area of the chimney at the bridge-wall, may cause more heat to impinge on that particular part of the boiler, but it will not cause the evolution of more heat. The sum total of the heat remaining the same, it will do the same duty, whether absorbed by a small part of the boiler, to which it may do injury, or by the whole surface at a more general temperature.

The extent of the sides of the furnace, when made

of brick, may be used as an argument against a large fire chamber; but the loss through a well-made brick wall by radiation is so little that it will not offset the benefit due to complete combustion.

Figs. 47, 48, and 49 show a horizontal multi-tubular boiler, as ordinarily set; 47 being longitudinal section, 48 half front and half cross section, and 49 floor plan.*

The different parts of boilers and their settings have technical names, applying to the corresponding parts of all boilers, as far as the construction will permit; the shape, sometimes, modifying the name, and increasing or lessening the parts. As an example, a *return-flue* boiler,

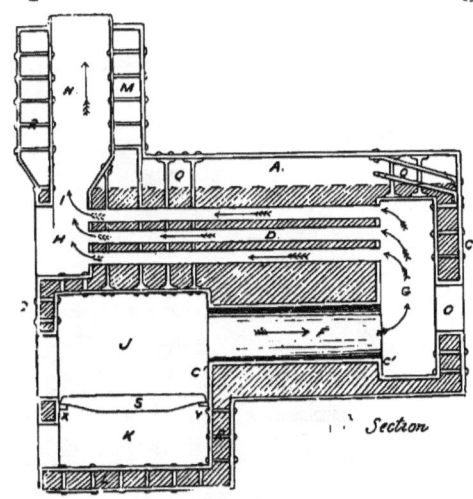

Fig. 32.

and a *drop-return-flue* boiler are shown. (Figs. 32 and 33).

The return flue boiler can be used as a stationary or marine boiler with or without a water-bottom; the drop-return being generally constructed for stationary boilers, as it has no steam chimney, and the smoke connection is a sheet-iron breeching.

* The proportions for boiler and setting, shown in Plate 2, are better than those just mentioned.

The following are the names or principal divisions of a boiler, and similar letters apply to similar parts in Figs. 32, 33, 47, 48 and 49:

A. Boiler-shell.
B. Steam-dome.
C. Boiler heads.
C'. Flue sheets.
D. Tube.
F. Flues.
G. Back connection.
H. Front   "   or smoke connection.
I. Smoke   "
J. Furnace, or fire-box.

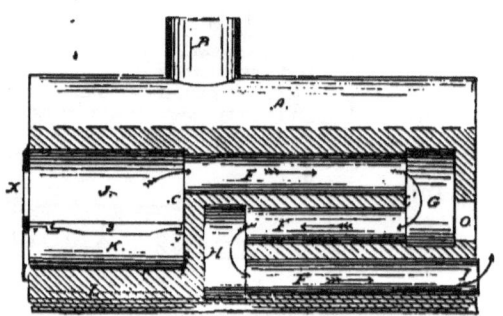

Fig. 33.

K. Ash-pit.
L. Water-bottom.
M. Steam chimney (marine).
N. Smoke chimney (marine).
O. Man-hole, to back connection.
P. Bridge-wall. (See Fig. 47.)
Q. Braces.
R. Stay, or socket bolts.
S. Grate bars.
T. Coking, or dead-plates.

U. Front-bearer. (See Fig. 47.)
V. Back-bearer. (See Fig. 47.)
W. Division, between front connection and fire-box. (Fig. 47).
X. Boiler-fronts, cast-iron.
Y. Side walls.
Z. Lugs.

The division between furnaces, and the sides of furnaces, are called "Legs" in fire-box boilers.

The same letters apply to the corresponding parts of the horizontal boilers, Figs. 47, 48, 49.

CHAPTER VI.

FORMS OF BOILERS USED IN HEATING.

THE conditions required for heating boilers, which are of such proportions that they may be fitted up to work automatically, are simplicity of construction, durability of parts, and ordinary economy in firing.

A source of danger to the success of the young steam-fitter and to others inexperienced in steam-fitting, is their endeavor to construct ideal boilers, which usually prove to be failures. It is far better to use boilers proved successful by others, and improve their weak points from experience with them. Success lies in that which will give *least trouble*, and will not wear out rapidly—the burning of a few tons of coal more or less in a year is not a proper test; as the conditions of management, the size of the house, the amount of ventilation, the number of hours the apparatus is operated in the year, and last, though not least, the comfort and satisfaction—all must be taken into consideration to prove economy.

Fig. 34 shows probably the simplest form of upright boiler used for heating, excepting, perhaps, one with a flat crown sheet. The grate is drawn in at the bottom, by a slanting annular dead plate, as shown; the center part of the grate only has openings. The brick-work is very simple, and is built around the boiler, leaving about a three-inch space for a flue, and the smoke pipe is taken out at the bottom. It does not rate very high in point of economy of fuel; but it is very easily kept clean, and lasts a long time. They are now seldom seen.

Fig 35 shows an upright boiler (multi-tubular), which is drawn in at the fire-box, to the size for the grate. This dispenses with the annular dead plate, and makes a very permanent piece of work. This boiler is set to carry the heat, when it leaves the tubes down one side of the boiler, and up the other, passing under a septum of iron, or a division wall, which may be run very near the boiler, but so as not to press against it. When the tubes of this boiler are not smaller than two and a half inches, or longer than three feet, and nothing but hard coal is used, it will require cleaning but once a year, provided there is no leak in the fire-box, or about the ends of the tubes.* To clean the boiler,—remove the cover

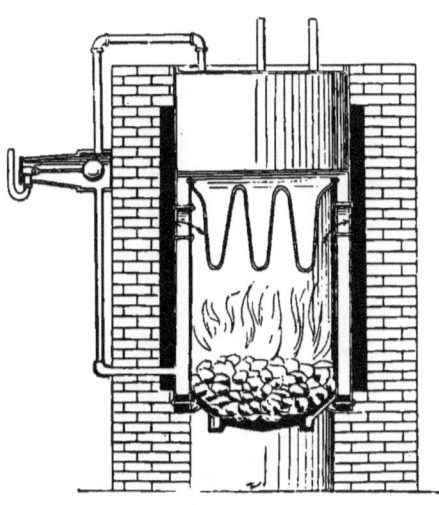

Fig. 34.

---

* Much moisture causes the fine white ash, which comes from hard coal, to bake on the heating surfaces, and should be prevented.

66   STEAM HEATING FOR BUILDINGS.

a', and use a steel wire tube brush. The cover a' is covered with abestos or magnesia on the top, and in the space c, around the top, to prevent radiation, or danger from fire. It will be noticed, this boiler is set on a cast-iron plate, to give it stability. This plate is more satisfactorily made in *two* parts, and bolted together, which will prevent the heat of the

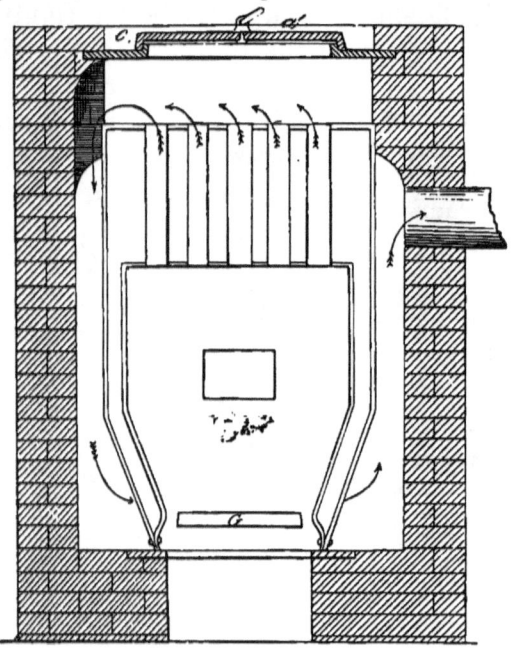

Fig 35.

fire from cracking it, after it is set. The grate is here shown, a little higher than it is usually set; but it would be well to keep it as high as the rivets.

Fig 36 shows the ordinary upright boiler, set for heating. It has a peculiar steam dome, as shown, which prevents an excessive heat on top, and it is claimed slightly superheats the steam. It also

FORMS OF BOILERS USED IN HEATING. 67

has an ash-sifting grate below the regular grate which saves much dust in the manipulating of ashes, and prevents the grate proper from burning out rapidly.

The form of dome shown here prevents the cleaning of the boiler tubes except with a steam blower. The

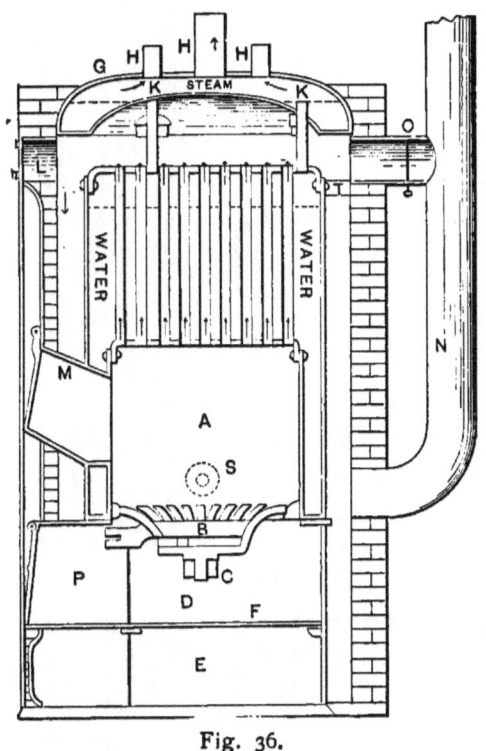

Fig. 36.

connections between boiler and dome have also to be of very large diameter, and circulating pipes—not shown—are necessary to take the condensation or water carried into the dome back to the boiler. When they are omitted the water carried into the dome is carried over into the heating pipes, and much noise in the apparatus is the result.

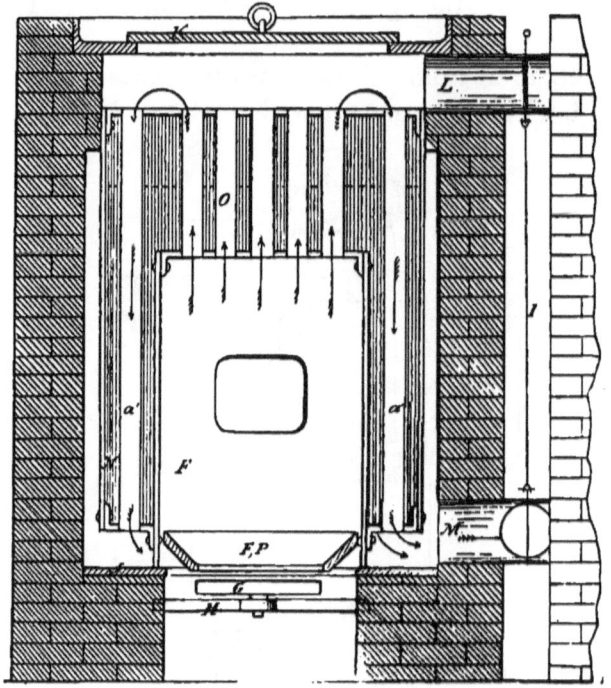

Fig. 37.

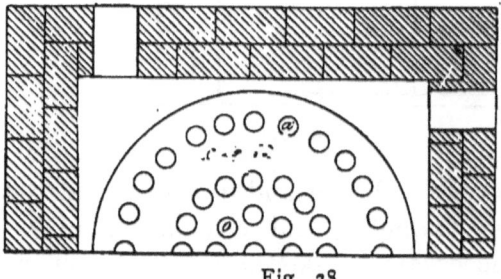

Fig. 38.

Figs. 37 and 38 show an *upright multi-tubular reverberatory tube boiler.* Fig. 37 is a vertical section on a center line, and Fig. 38 a half cross-section, to show the walls and tubes. In Fig. 37, FP is the fire pot, or dead plate; F, the fire-box or furnace; G, the grate; H, a bar set in the brickwork of the ash-pit in such a way it may be removed to put in a new grate, and into which the grate is pivoted, a certain distance below the edge of the fire-pot, to admit of shaking and cleaning from the bottom. The amount of opening is regulated by washers on the pivot of the grate, to suit the size of coal used; O, the direct tubes; a', the reverberatory tubes; J, the bottom plate; K, the cover; L, the direct chimney flue; M, the bottom or drop chimney flue.

In point of economy of fuel, probably there is no house-heating boiler stands higher than this, if properly proportioned, and in permanency it is fully equal to any wrought-iron boiler used; besides, it is not difficult to clean. It will be seen that all the flues are internal, and if the gases of combustion cannot impart any heat *to* the boiler, after cooling to a certain degree, they cannot abstract any *from* it, as happens in external flues, when the gases cool to the temperature of the steam or below it, by an admixture of air through the brickwork before reaching the chimney.

It is also an excellent boiler where light power is desired, in which case the tubes may be of smaller diameter than would be used for heating, and longer, to suit a higher rate of combustion.

When upright boilers are enclosed in brickwork, the outside is usually built square, to suit the door castings, and for appearance; but the inside is generally built *round*, three or four inches from the boiler, to make a flue or an air space, which will be the same distance from the boiler at every part. A wall so built generally cracks in the thinnest part, which makes it necessary to build the wall square inside and outside, as shown with cleaning doors at

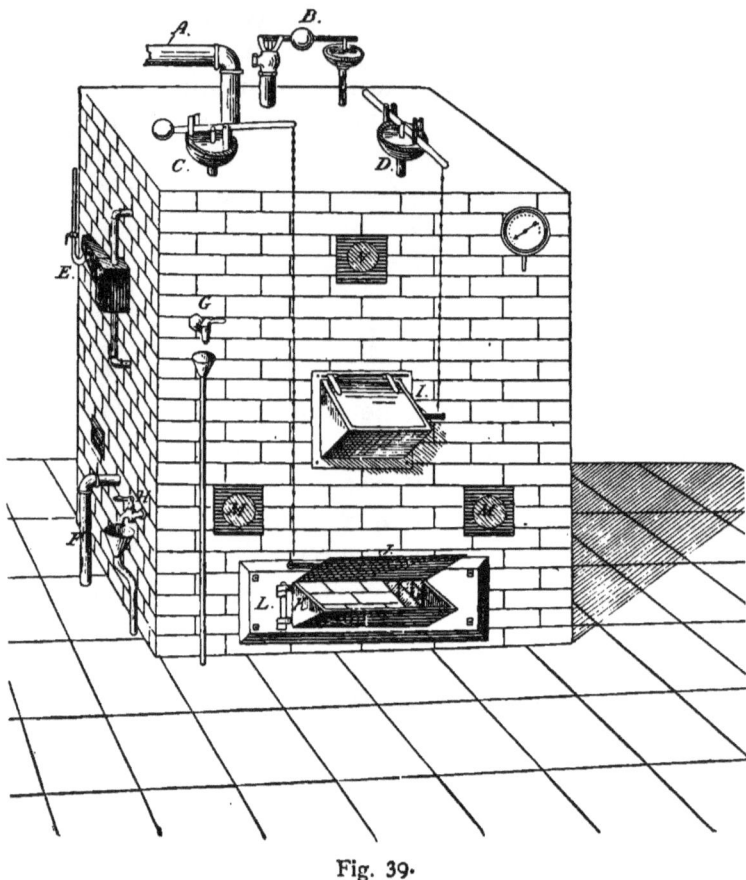

Fig. 39.

corners. The infiltration of air through the walls of brick-set boilers is a great source of loss.

When wrought-iron boilers are constructed for low-pressure heating, have them built just the same as if they were intended to carry high steam, taking care the leg, the part formed by the side of the fire-box, and the shell, is properly stayed with socket-bolts, or stay-bolts, as boiler-makers often show a disposition to leave the legs unstayed, when they know the boiler is for very low pressure.

Fig. 39 represents this boiler when set and fully fitted with the necessary self-acting appurtenances. A is the main steam pipe, which must be run for no other purpose but to distribute steam to the heaters ; B, the safety valve, with its auxiliary diaphragm ; C, the draft-door regulator (the pipe carried up inside the brickwork); D, the fire-door regulator, which is not absolutely necessary; but it is well to have, in case anything should prevent the draft-door from closing ; E, the automatic water regulator, whose connections should not be a branch, from any other pipe —nor should they be branched for any purpose ; F, the main return pipe, which should have no valves in it, unless there are valves in the main steam pipe to correspond. When there is but one boiler, it is generally better to dispense with valves in steam and return pipes at the boiler. G, the gauge cock, which for cleanliness may have a drip-pan under it, connected with the ash-pan ; H, the blow-off cock, which in a heating apparatus *should never be connected directly with the sewer or drain*, but should be a lever handle cock over a tunnel, as shown, to prevent the *possibility* of water passing out of the boiler without the knowledge of the person in charge. The tunnel can be removed when not in use. I, the fire-door, on a good slant, so as to form a shute for the coal, and to close without a latch ; J, the draft-door, an attachment to the ash-door; *K*, the ash-door, which is hinged to the frame *L*, and will open without interfering with the draft-door; the chain and the bolt having nearly the same *common axis;* *L*, the ash-door frame, which is bolted to a skeleton frame, built into the brick work, that can be removed to put in a new grate; *M M*, are hand holes, to clean the space at the bottom of the drop tubes; *N*, a hand hole, to clean the upper tube

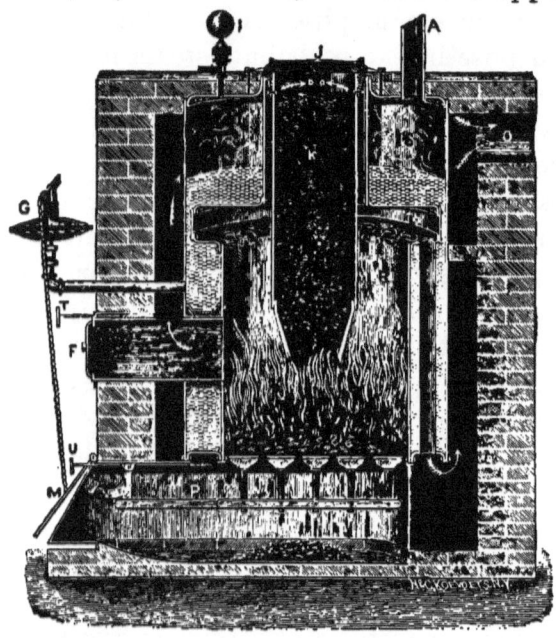

Fig. 40.

sheet, and through which a steam tube cleaner may be used, if desired.

Fig. 40 represents a wrought-iron boiler, which came into public notice about twenty years ago, and has given good satisfaction.*

---

* It was patented by Mr. Wm. B. Dunning, of Geneva, N. Y.

It is a reverberatory *tube* boiler, with a *coal magazine*, similar to the base burning stoves, and is entirely constructed of wrought iron, except the cast-iron magazine. When set, according to the manufacturer's instructions, every part of the boiler is exposed as heating surface; the heat passes between the magazine and the fire-box, and thence down the drop tubes, D, and up and around the shell. The magazine is made to pull out, and care should be taken when setting them, to have sufficient room overhead to accomplish this.

The heating boilers I have so far shown are all of wrought iron types. Since first writing this book, however, many cast iron house heating boilers have appeared in the market.

Presumably the first cast iron sectional boiler to make any appreciable headway and to remain permanently in the market is the "Mills" Boiler, made by The H. B. Smith Co., of New York. It consists of a number of cast iron sections A A, such as shown in Fig. 41, joined to a steam drum B, and to two water headers B' and B' by locknut nipples. The sections A A are practically upright tubular units, two of which when put together in the manner shown, form what is called a section, their depth being about six inches. A number of these sections are added together to form the boiler. The sides of the ash pit P may be formed of masonry so as to form flues F F with the outside brickwork C. The sections are then built together on plates covering these flues, which form a foundation. The grate line is at G. The direction of the fire, therefore, is backwards and upwards from the grate, returning to the front

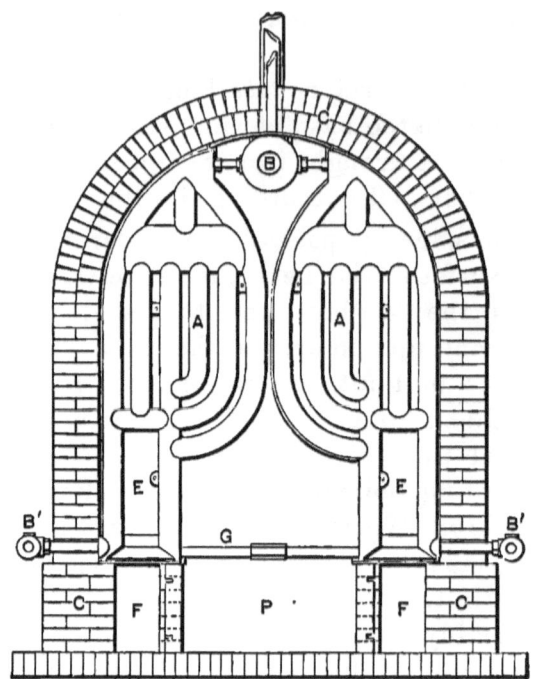

Fig. 41.

73 a Fig. 42.

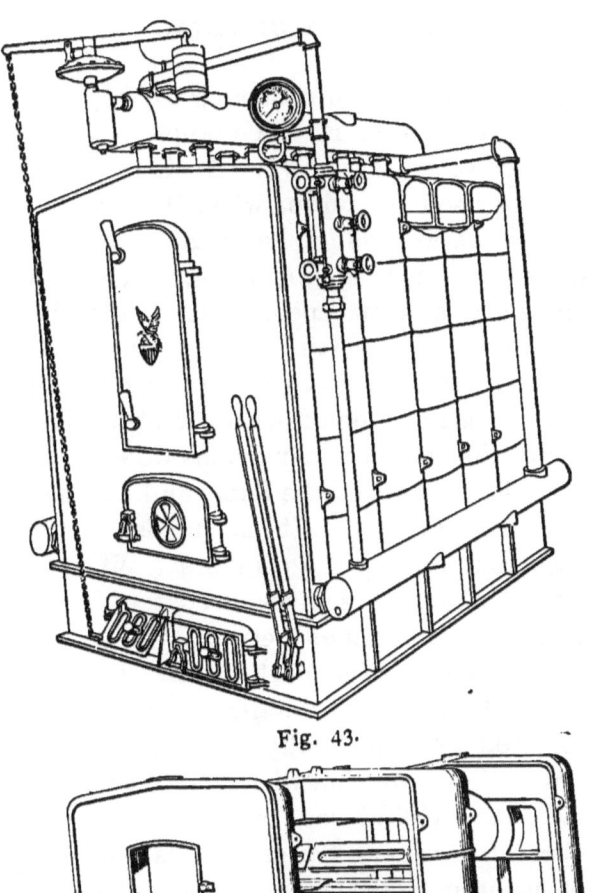

Fig. 43.

Fig. 43a.

through the flues E and again returning to the rear of the boiler through the flues F. The boiler is usually enclosed in brickwork as shown. It is a type of boiler that can be used for power when the pressures are not very high, as well as for heating.

The "Mercer" boiler, made by the same company, shown in Fig. 42, is a later type of this boiler which does not require a brick setting. As will be seen, there is a special front section with fire box sections and a rear section all connected with the steam drum on top, and with the return water drums on each side of the bottom. The course of the flame and flue gases are shown by the arrows. It is furnished with shaking grates that cut up the clinkers. When the boiler is put together, it receives a coat of plastic asbestos cement which clinches between the tee bars T, making a permanent and smooth finish. The illustration is so good that other details of construction can be understood without further explanation. An earlier type of boiler somewhat like this, but which is not shown, was the "Gold" cast iron boiler.

Another type of sectional boiler that is connected with headers in the manner just described is the Gurney, shown in Fig. 43 and 43a. The general outside appearance of the boiler is like that shown in Fig. 42, but inside it differs very materially from all other cast iron boilers by having its principal heating surface composed of horizontal circulating loops. A loop very similar to the Bundy radiator loop is screwed into the intermediate sections, as shown in the centre of Fig. 43a, giving a very large quantity of heating surface in a comparatively small space. A modification of the same boiler is made circular in two sec-

FORMS IN BOILERS USED IN HEATING. 75

tions. They are made by the Gurney Heater Mfg. Co., of Boston.

As an illustration of what may be done in a single casting or almost a single casting, we show the boilers, Figs. 44 and 45. Fig. 44 is a boiler made in two parts, known as the Cottage, and made by The H. B. Smith Co., of New York. Fig. 44 shows the several sections and also the arrangement of the parts, including the

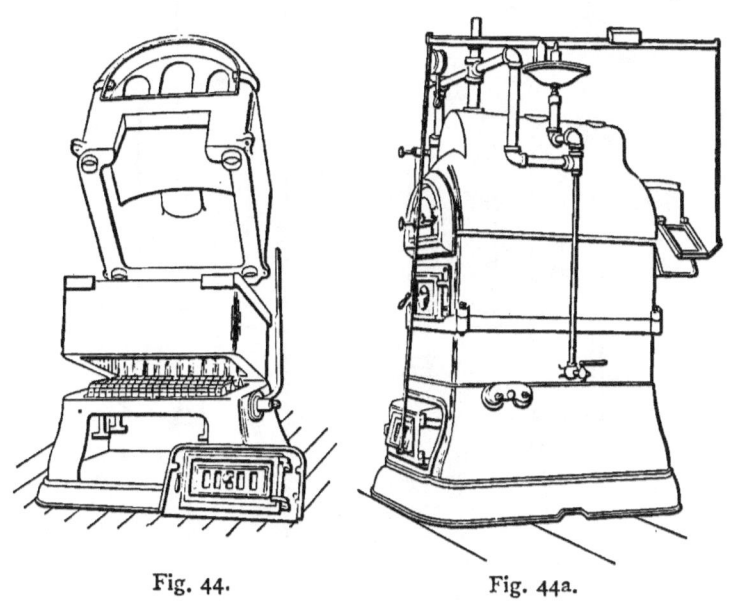

Fig. 44.   Fig. 44a.

grates, fire surface and flue passages. The upper part forms the boiler proper made in a single casting. The flue gases pass backwards and upwards and forward through the two side flues, and return to the rear again through the centre flue. The mid-section shown in the illustration, forms a water leg about the fire, and the boiler proper and the water leg are joined together

76   STEAM HEATING FOR BUILDINGS.

by slip nipples. Fig. 44a shows the general appearance of the boiler when set up and used for steam.

Fig. 45 shows a vertical water tube boiler made by the Gurney Heater Mfg. Co., of Boston, which may be

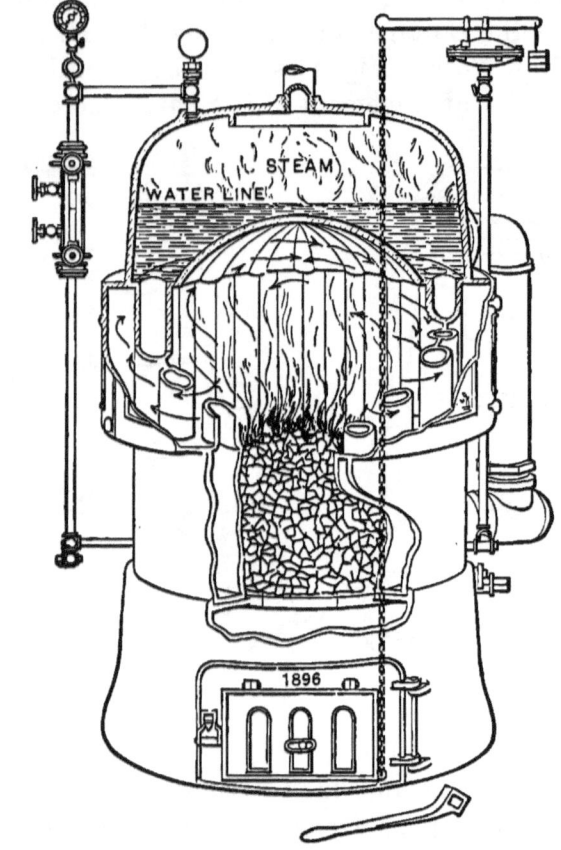

Fig. 45.—(The Doric.)

said to be made entirely of a single casting, and is so made so far as the water and steam parts of the boiler are concerned. The illustration shows the arrangement of the boiler so thoroughly that comment is un-

FORMS OF BOILERS USED IN HEATING. 77

necessary. The Cottage and the Doric boilers, of course, are made for comparatively small heating apparatus, while the other types of cast iron boilers run up to very large sizes and can be used in batteries for the very largest description of work.

Fig. 46 shows a recent type of cast iron sectional boiler put on the market by the A. A. Griffing Iron Co., of New York. Its general difference from other

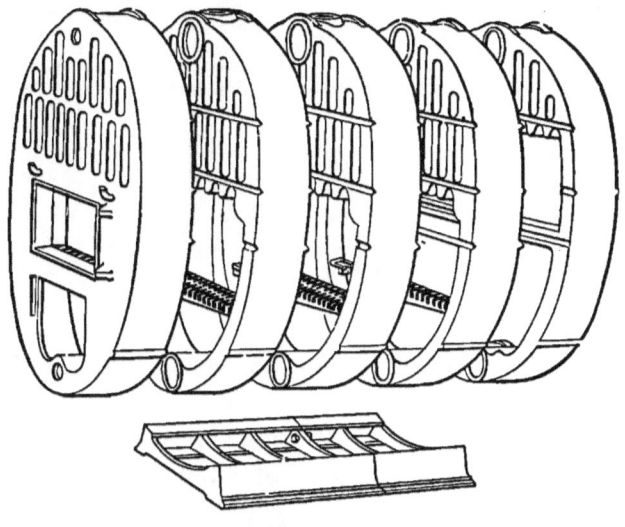

Fig. 46.

boilers of its class lies in the fact that it is put together entirely by slip nipples, the ash pit being formed by the lower part of the section proper; the sections themselves simply setting in an iron cradle on the foundation. The crown sheet of he firebox is corrugated. The gases of combustion pass forward and backward through the tubes, the upper row being superheaters. These boilers, I am informed, are being used for both heating and power. They are called the "Bundy."

THE HORIZONTAL MULTI-TUBULAR BOILER.

Figs. 47, 48 and 49 show longitudinal section half front elevation and half-cross section, and plan,

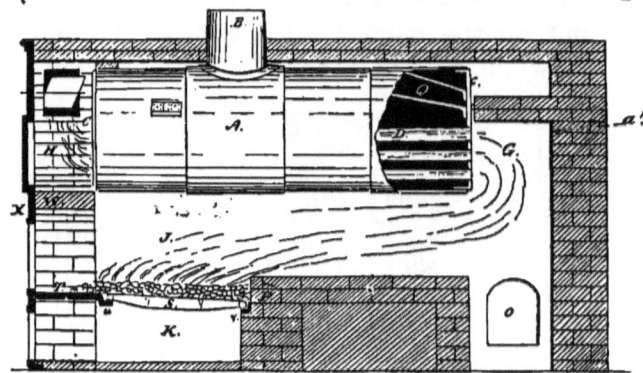

Fig. 47.

of an ordinary horizontal boiler, set for heating or for power.

This is the style of boiler most in use in the United States, when the building is of such proportions that

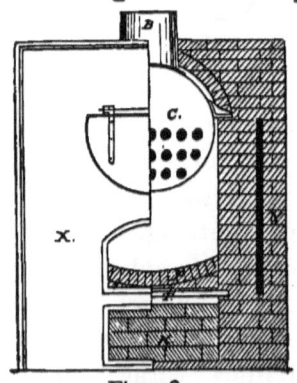

Fig. 48.

it requires power, and considerable notice will be given to it—its method of construction, setting and so forth, as it is the typical American boiler. They are sometimes fitted with automatic appurtenances, but where two or more of them are in a building, automatic draft regulators are all that should be used; and a careful engineer or fireman should do the rest.

When used for power where the water contains mud, as in some western cities, they should be fitted with a mud pipe, as shown in Fig. 50, or if used for heat-

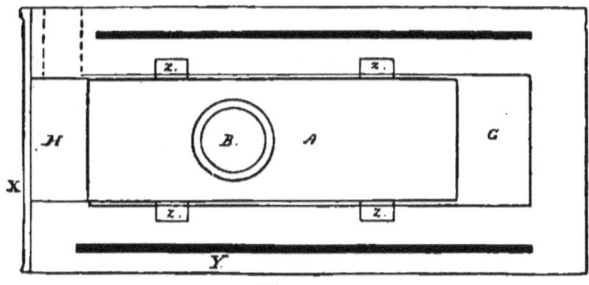

Fig. 49.

ing when the water is wasted; but this is scarcely necessary in a gravity apparatus.

Fig. 50 shows a horizontal boiler where the front end of the shell is supported by resting in the cast-iron front; with the front connection formed by what is known as *breeching;* this is sometimes made of light iron and bolted on; but it is better to form it by an extension of the boiler shell, as shown. This dispenses with the division *W*, as shown in Fig. 47.

There seems to be a dislike to this front, for no better reason than because it is not considered ornamental. It is certainly a substantial front, if made in sections and bolted about the doors, where all fronts are liable to crack, and if set as shown with deep dead-plate and two courses of firebrick lining, it will seldom require repairs; but if the front bearer is bolted to the cast front, and the front is lined with a single course of fire-bricks, held in their place around the door by a cast-iron frame, the frame will burn off, the

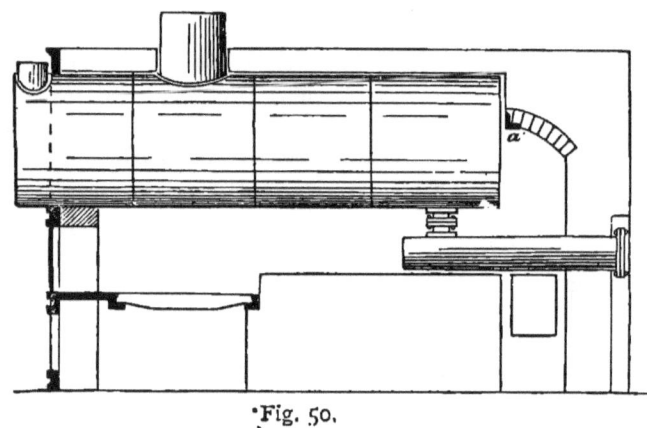

Fig. 50.

lining fall down, and the front become heated and cracked. With a straight or "flush" front, a dead plate is always used, to keep the fire away from the front connection W, Fig. 47. The thickness of the wall necessary to form the front connection forms a lining for the front, which *must* be kept in repair, or W will fall, and as W cannot be dispensed with in a boiler set, as in Fig. 47, the front is thus preserved. If the dead plate is used and made sufficiently deep, whether W is used or not, the front will last!

This front and setting also obviates the necessity for the projection, shown in Fig. 47, which is spoken of elsewhere.

PLATE 2.

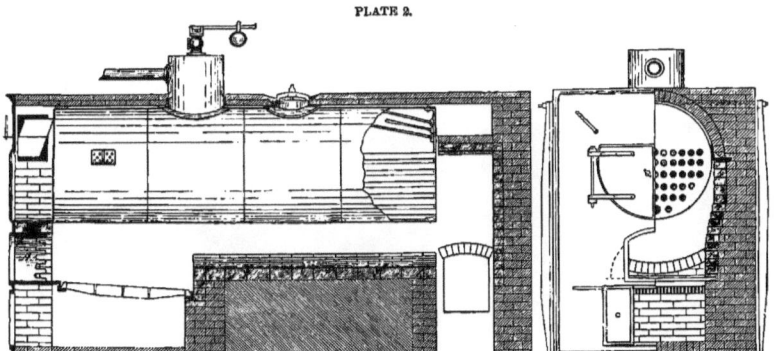

HORIZONTAL-MULTI-TUBULAR-BOILER.

Plate 2 shows a horizontal multi-tubular boiler, similar to the boiler shown in Fig. 47, but with the *improved cast-iron fire-door arch A;* with the manhole on the shell, domed steam drum, flat gusset braces, and other details of a modern steel boiler.

It was usual to make the shells of No. 1 charcoal hammered iron—though nearly all are now made of a fine grade of boiler steel. When steel is used, shells up to 42 inches should be made of $\frac{1}{4}$ inch plate; from 42" to 48" of $\frac{5}{16}$ thick plate, and from 48" to 60" of $\frac{3}{8}$ to $\frac{7}{16}$ thick plate; with head sheets of $\frac{3}{8}$ to $\frac{7}{16}$ and $\frac{1}{2}$ respectively, shells and heads being constructed of best flange steel.

The domes of these boilers are usually made one-half the diameter of the shells, and about the same height; but the limited height of cellars often reduces the height of the dome, and in some cases renders it necessary to dispense with them altogether.

The height for the setting of a 48-inch shell should not be less than 11 feet, and as much more as can be conveniently had. This will allow 2 feet from the paving of the ash-pit to the grate, and 2 feet more from the grate to the boiler, 4 feet for the boiler and 2 feet for the dome, leaving 1 foot from top of dome to underside of sidewalk or floor beams. For each additional foot of diameter of boiler 16 inches should be allowed.

Low cellars are a detriment to a heating apparatus in another and very important respect—they bring the main steam pipe too near the water line of the boiler, and make the use of mechanical devices necessary in work which otherwise could be made more perfect as a gravity apparatus.

When the man-hole of a boiler is in the top of the dome, a hole in the shell underneath the dome, large enough to easily admit a man from the dome into the shell is required. This is bad practice, as this large hole weakens the boiler materially; which fact engineers generally pay no attention to. The shell of a boiler underneath the dome should not be cut out unless it is reinforced in some proper manner; but should be perforated with a number of small holes—say 2 inches in diameter—until their aggregate area is four or six times that of the steam pipes.

When the man-hole is in the top of the boiler an extra heavy man-hole frame should be riveted to the shell; its longest diameter being *across* the shell.

The tubes in horizontal boilers give the best results when not "staggered," but placed in vertical rows and should have at least *one* inch between the tubes at their nearest parts, and should be not nearer the shell than 3 inches.

These boilers should be tested to 150 lbs. per square inch by hydraulic pressure. This is absolutely necessary to test the bracing and other parts, such as heads and man-hole frames.

There is a prevalent idea that testing a boiler with cold water may injure it. If a boiler will not stand twice the ordinary pressure it is made to carry without injury under a hydrostatic test, with water at 40 degrees Fahrenheit, *it should not be put into a building*, and the constructor or engineer who makes such an assertion does so either through ignorance or through the fear that his apparatus is not up to a reasonable standard of strength.

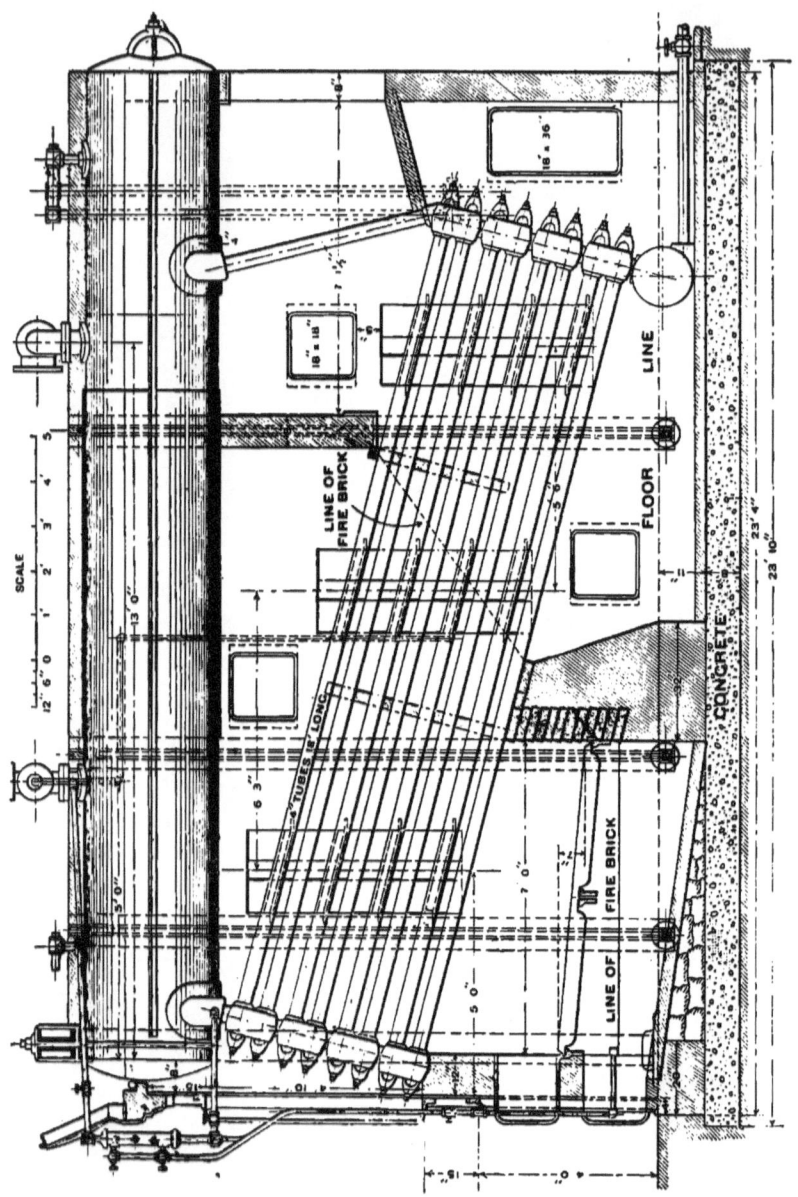

Plate 3 shows a water-tube boiler. It is of a class that may be used for either heating or power. It is known as the Caldwell boiler, made by James Beggs & Co., of New York, and represents an improved form of the Babcock & Wilcox type. This boiler differs from the regular Babcock & Wilcox in the formation of its headers. In the latter the headers are of one piece from top to bottom of each section, while in the Caldwell the headers are made up of four or five short elements, each element having but four tubes expanded into it. Several of the elements are then put together, forming a section of the boiler. The handhole plates in the Caldwell boiler are also on the inside of the headers, so the pressure does not come on the bolts, but on the gaskets. The heights of these boilers usually prevented their use in basements or cellars with a gravity apparatus, but with a modern pump-governor apparatus they are fast coming into general use, principally on account of their comparative safety. Now modified forms are being made, called "architects' boilers," which are lower, that are intended to overcome this difficulty.

Fig. 51 shows a Root water-tube boiler, manufactured by the Abendroth & Root Mfg. Co. of New York. It is a very early type of the water-tube boiler, and from its earliest form the Babcock & Wilcox type was probably developed. Its present form is made up of a number of two-pipe elements, the elements being arranged in vertical staggered rows, and connected back and front with special bends and a peculiar form of joint. Each vertical row connects into a steam- and water-drum, so that five, more or less, of these rows or larger elements make up a boiler, the drum of each large element connecting with the steam-drum. Circulating pipes connect between the small drums and the back headers of the boilers so as to give the water a circulation through the smaller drums. The large cross or steam drum is also connected to the rear cross headers by a dry-pipe, so that any water carried over by the force of ebullition or otherwise is returned to the lower part of the boiler from the upper or steam drum.

The method of setting and the other general form of parts is plainly shown in the illustration.

Fig. 51.

A point in the favor of water-tube boilers is their greater safety when compared with the shell boiler. Eleven feet in these boilers is rated as a horse-power, while fifteen is the usual rate in the multi-tubular shell.

CHAPTER VII.

GENERAL REMARKS ON BOILER SETTING AND CONSTRUCTION.

THE best materials should be used in the settings of boilers, and less than a 12-inch wall should not be allowed even in the setting of the *smallest class of horizontal boilers*. Large boilers should have not less than 12-inch walls in addition to the thickness of the fire-brick lining of the furnace, and 20, and 24-inch walls are not uncommon.

It is not desirable to put a number of masons on boiler walls and hurry them; for neatness and deliberation are required with every brick, and makeshifts should never be allowed.

On marshy, or sandy ground, it is well to excavate for the whole size of the apparatus and put in a thick concrete foundation, which will keep the work substantial and also help to cut off moisture from the earth.

It is generally assumed that the greater expansion of the bricks on the inside of the furnace is the cause of the boiler walls cracking; and it is, to a large extent true, though cracks from this cause are generally distributed over the walls, and are not so great but that a few coats of whitewash are sufficient to fill them.

The large fissures which often appear in sidewalls of boilers are usually caused by an insufficient foundation, the walls resting on or against the boiler; or by unequal or abrupt changes of thickness. Opposite the bridge-wall a crack usually appears which is supposed to be caused by the mass of the bridge-wall moving in a different direction to that of the wall. This crack nearly always appears, and as it opens under the heat, small particles sift down within it and prevent its closing when cold, and this action going on often opens a large fissure.

The arch over the back-connection of a boiler should not be turned against the boiler-head, as is often the case, but should be sprung from the side walls; with a rod to form the chord of the arch with the necessary flanges or buck staves on its ends, the rod to be just covered from the heat in the back wall.

If it is desirable to turn the arch from the back wall to the back head of the boiler (since some think this shape more desirable), a heavy angle iron should be used to turn the arch against. The angle iron should be kept half an inch from the boiler, taking care no mortar or bits of brick lodge between the head of the boiler and the angle iron.

When the lugs of a boiler are firmly built into the brickwork, without iron plates in the wall for the lugs to "give and take" on, the walls will crack, because the iron of the boiler contracts and expands more than the wall does. The lugs should also be free from the brickwork on their ends and top.

The arch, turned over a boiler, should not touch it, but there should be one or two inches of space between boiler and arch. The arch should spring from the side walls, and be self-supporting, and not turn on the boiler.

A good way to build these arches is to lay inch strips of wood lengthwise on the boiler and draw them out as the work progresses.

When boilers are not arched over, but the sidewalls are run straight up, and the space, over the boiler, filled with sand or loose materials, the walls are very apt to crack and be shoved out of plumb. Every time the boiler cools, the sand and loose particles will press down between the boiler and the wall, and the whole mass above will settle down. Then the boiler becomes heated again and expands; the sand will not be forced up again; hence the wall will be shoved out. This often happens, and it is attributed directly to the action of the heat, as something unavoidable, but such is usually not the case.

When boilers are set on sandy ground the foundation should be deep and good, as the heat of the furnace will drive out the moisture from the sand and leave it a *quicksand*, so that they should have a heavy foundation of concrete that the whole mass may settle as a monolith.

An air space within a boiler wall is of doubtful utility, the same thickness of brick will prove more serviceable and will not weaken the wall.*

The fire-bricks in a furnace, should have the smallest possible quantity of fire-clay between them, barely sufficient to level the work. They should be

---

* I do not wish to convey the idea that a space in the walls of a building is not valuable; since it interrupts the passage of moisture, the evaporation of which, from the walls, would require more heat than would be lost otherwise.

laid with a couple of courses of headers at the top, so the side linings could be removed without affecting the stability of the wall. The other courses should not have headers, but an occasional header to tie the face of the wall, as the breaking out of a row of headers will injure the structure of the wall.

The division (W, Fig. 47) between the furnace and the front-connection is another source of annoyance; when constructed of iron it burns out rapidly, and when made of fire-brick, in the shape of an arch, it falls out; or may be broken in using the fire-tools.

Hollow castings, with air and water circulations in them, have been tried to form the under side of the front connection, but they do not last and are dangerous under high steam pressure, as they are usually flat-sided. The shell of the boiler is sometimes allowed to project and cover this space; but as it has heat on both sides of it, it buckles and burns out in a year or so, unless the engineer is very careful about keeping the brickwork in order.

Sometimes the shell is extended with a water space, formed on it by a projection of the head sheet and shell, which forms a permanent fixture; and if the part is well studded with stay-bolts there can be no objection to it; but care must be taken, when a high pressure of steam is to be used, as this "shovel nose" (the name by which it is known) will form the weakest part of the boiler (see Fig 52).

If an iron arch is used underneath a brick arch to support it and keep it from being knocked out, it will last longer; but the inner edge of the casting will bulge and get out of shape long before the iron will be burned away, which suggested to the writer, that if the cast-iron arch (which should spring from the dead plate and form the doorway to the furnace) flared inward, and was cut into, for about one third its depth, making large and coarse prongs (about 2 inches wide by six inches long, with one inch of a slot) to support and guard the bricks, it would stand

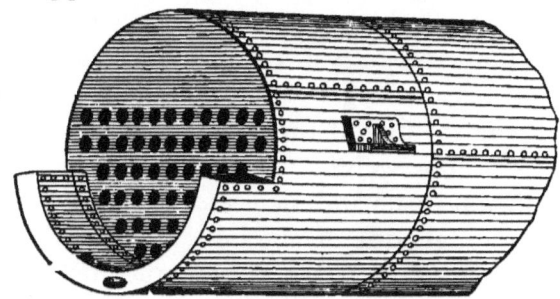

Fig. 52.

for a longer time. This method has been used many years, and the prongs do not bend down, while they burn off very slowly from their points, lasting three or four times as long as the ordinary cast-iron door frame.

A deep dead-plate saves the front and door linings, as it keeps a body of comparatively dead coals between the front and the fire.

Bridge walls are often built straight across, but an inverted arch is better; though not on account of combustion, but that in an arch the bricks are keyed in, and are not as likely to be knocked out by the fire tools.

Deep ash-pits are the best, and a second or ash-grate will help preserve the grate-proper; as there is less reflection of heat from it than there would be from a hard brick bottom.

The brackets riveted to sides of boilers to support them in the brickwork are commonly called "lugs," and many engineers, in the construction of what they consider long boilers, put *three* lugs on a side, fearing the weight will be too great for two only. This is undoubtedly a mistake, and frustrates the object for which the third is put on. The object of the extra lugs is to distribute and lessen the weight on any one lug. With a middle pair of lugs, however, the settling of the brickwork at one end will throw the whole weight of the boiler on the *middle pair*, and even if the walls should not settle, the heating of the under side of the boiler more rapidly than the top, which takes place for instance upon starting a fire before steam is up, will in a great measure force up the ends of the boiler, leaving the whole weight on the middle pair of lugs.

Four lugs, properly put on, are found to be the best number, and the detail, Figs. 53 and 54, page 92, shows a method introduced by the writer and now very much in use by careful boiler-makers. The lug is extended downwards, the part going against the side of the boiler covering a greater area of the boiler shell than usual, and bringing a row of rivets below the horizontal bracket of the lug. With the old-fashioned lug all the rivets are above the bracket, and the tendency of the weight is to put an undue strain on the rivets, which to many is supposed to be a shearing force, but which in reality is very much more of a pull in the direction of the length of the rivet, the tendency being to pull the rivet through the side of the boiler or break the rivet in its shank. When the lug is extended below the bracket, as shown in Fig. 53, the tendency to pull the rivet through the sheet is greatly lessened; and, in fact, if the lug is properly proportioned, the strain on the rivet becomes almost entirely a shearing strain, the resultant force being almost directly downwards, or the tendency of the bracket to tear off, almost directly upwards.

It is a mistake also in setting boilers to neglect to get the support under the bracket close to the side of the boiler, because when the bracket bears unevenly, or near its outer end, on its roller, it increases the leverage to tear the bracket from the shell. Many initial ruptures in horizontal boilers occur underneath this bracket, and for this reason many makers of large and heavy boilers are abandoning the bracket and using the method shown in **Plate 7,** and one that has been used a great deal on river steamers in our western waters. Flat, wrought-iron or steel suspenders are riveted to the side of the boiler, which is supported on I-beams, crossing from wall to wall, as shown. With this style of support there is very little danger of tearing the suspender or "lug" from the side of the boiler, as the strain on the rivet is entirely a shearing strain, but there is danger of getting the pins or rods that run to the beams above too light; and in designing the rods and suspenders care must be taken that any two opposite suspenders will sustain the maximum load with safety, the maximum load, of course, being the weight of the boiler when it is full of water.

Lugs are sometimes left off until a boiler is in the basement, for the purpose of getting it through doorways. This is not good practice, as the rivets should be driven on the line inside of the shell, before the tubes are put in. Putting them on with tap bolts is not good practice either, as two or three bolts may have to carry the whole end of the boiler. Bolts of $\frac{3}{4}''$ or $\frac{7}{8}''$ diameter tapped into the side of a boiler, and loaded as a bracket bolt will be, are apt to break or strip in the thread, and there is no way by which the boiler maker can safely ascertain the load on a bolt. He is obliged to strain on the bolt until he draws the work close, at which time the bolt often breaks, or is ready to break, so that there is no factor for safety that can be depended upon. Should the bolts leak also under pressure, the brickwork has to be torn down to remedy the defect, and the work is often made tight with cotton wick or some similar packing.

A good plan when the lugs must be left off, is to have a shoe riveted to the boiler at the proper time, into which the lugs will slip, similarly to a stove leg, and which, of course, must be sufficiently strong for the work.

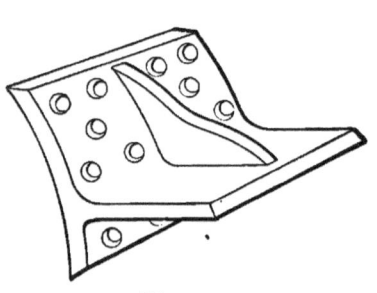

Fig. 53.

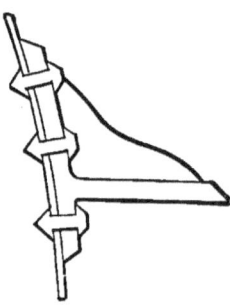

Fig. 54.

CHAPTER VIII.

PROPORTIONS OF THE HEATING SURFACES OF BOILERS TO THE RADIATING SURFACES OF BUILDINGS.

THERE is no simple relation between the heating surfaces of boilers and the radiating surfaces of the buildings they have to supply the steam to, as the following considerations modify every type of apparatus: The class of boiler used; the method of setting boilers; what the grate surface is; the character of the work the boilers are designed for, and whether the air is simply to be maintained at a certain temperature, as in direct radiation, or whether every cubic foot of air which comes in contact with the radiator must be warmed from the outside to the inside temperature, as in indirect radiation, or whether the apparatus is *direct-indirect* or composite. All these will have to be considered, and the results are then only close approximations to the truth. Neglect of cleaning, a certain amount of neglect of management, and the state of the fire—whether on the first hour of the new fire, or the last hour of the dirty fire—for the time they are to run without attendance, all must enter into this calculation, and then one is generally forced to err on the side of safety—that is, not have just sufficient boiler to do the work, but a little more than enough to do the work under the poorest conditions likely to be encountered.

If the effect of the cooling produced by loss of heat through the glass and walls of a building can be properly estimated and added to the amount of heat lost in warming the air admitted for ventilation, a close estimate can be made of the smallest grate that will burn sufficient fuel to evaporate the required amount of water in a boiler *sufficiently* large, but not so large as to be wasteful. The following points the constructor must also keep in mind in estimating for an automatically fitted boiler, that it is the amount of opening of the draft door or damper which regulates the fuel burned; the fuel burned regulates the water evaporated; and, finally, the water evaporated regulates the heat—both the heat of the room, by supplying steam to sufficiently large heaters, and the heat or pressure necessary to move the diaphragm, which in turn regulates the draft, and that the amount of steam made, so that really what is required are certain limits, within which an engineer knows he is safe, and to exceed which would be an unnecessary expense.

Boilers for very large buildings, which have an engineer in charge, should be figured pretty closely, as he is supposed to be constantly at his post and to clean his boiler fires regularly, and to fire often and in small quantities; keeping his fire door open the shortest time possible, and further, to clean the tubes or flues whenever required. But this is not the case in house boilers, for they must run for long periods without cleaning or interruption, and be adequate to every contingency of change within their limit of time to keep steam without attendance.

It has been found by experiment in a general way, and from practice, that for ordinary large buildings, with average window surface, and for the greatest range of temperature in our northern states, when nothing but direct radiation with no ventilation is used, one square foot of boiler to every ten square feet of the radiating surface will answer; assuming, of course, the radiating surface is ample. This is an approximation for high pressure boilers only with fair to good draft, and though I have one case in New York where one square foot of average boiler surface supplies from 14 to 15 square feet of radiation, there are many boilers under the ordinary condition of setting with short or cramped chimneys that will not do better than 1 to 8, and this for low pressure steam only (2 to 10 pounds pressure), as, of course, the higher the steam pressure the greater the condensation, a matter that will be referred to hereafter.

For indirect radiation, if the heating or radiating surface of the coils are *double* what they would be for direct radiation without ventilation, the *same* proportion of boiler to coil will about suffice; but, if instead of doubling, the same surface is used or a slight increase or even decrease is used and the building is kept warm by moving the air faster as with a fan, through a comparatively small coil, we must proportion the boiler the same as if we had double the surface, that is, 1 of boiler to 5 of indirect radiator. These rules are only the roughest of approximations and often lead to much blundering, and if one will only bear in mind that a boiler should be proportioned to the cooling which goes on—heating, ventilating, etc.,—and not to the coil surface, as that is as variable as boilers themselves, they will not expect a direct answer to their question.

For direct-indirect radiation, proportion the boiler about one and one half times what it would be for direct radiation.

These estimates are for boilers with ordinary high combustion, such as horizontal boilers which are kept clean without interruption; but for house boilers with slower combustion, an addition of $\frac{1}{4}$ to $\frac{1}{2}$, depending on the type of boiler and good judgment, will be required on the part of the engineer.

The manufacturers of the boiler shown in Fig. 34, make 3 sizes, of 45, 60, and 75 square feet of heating surface, and say they will furnish steam for 300, 500, and 700 square feet of direct radiation coils. These boilers were used for many years in the early days of steam heating, and were probably not over-rated. They are of a simple type with direct surface, but wherein the gases of consumption escaped into the chimney at a high temperature. They were not as economical of fuel as more modern boilers, but illustrate a type in which the proportion is about *one* of boiler to 9 of radiating surface.

The manufacturer of the upright tubular boiler, shown in Fig. 36, published a list of 24 sizes of boilers, from 54 square feet of surface to 400 square feet, in which he gives the maximum and minimum number of cubic feet of air in ordinary buildings each boiler will carry radiation for.

The following is a condensed table of this list:

No. of Boiler.	Feet of Surface of Boiler.	Maximum and Minimum of Cubic Feet of Air in Building.	Square Feet of Radiation.
1	54	18 to 25 thousand	360 to 500
6	107	40 " 54 "	800 " 1080
9	151	55 " 75 "	1100 " 1500
12	202	72 " 100 "	1500 " 2000
18	302	116 " 152 "	2320 " 3040
24	403	164 " 215 "	3280 " 4300

There is no doubt this list is approximately correct when upright multi-tubular boilers are used, or any kind of *shell* boilers, with simple parts. The proportion runs between 1 to 7½, and 1 to 10.

In the Nason Manufacturing Company's old catalogues on thin pipe boilers, a circular is to be found giving the following list—in which the grate to the heating surface of the boiler is about as one to 27, and the heating surface of the boiler to the radiating surface of the building 1 to 6½. This was considered a safe and liberal allowance, which it proved to be, as under favorable conditions ⅛ to ½ more direct radiator surface would be carried by the boilers.

Square feet of Grate Surface............	2	2½	3	3½	4	4½	5	6	7
Square feet of Boiler Surface exposed to the fire............	55	65	78	83	105	116	131	158	182
Square feet of Radiating Surface which it will heat............	350	440	525	600	700	775	900	1050	1225

The proportion of grate surface shown here also gave good practical results.

Morris, Tasker & Co., of Philadelphia, give a list in

which the rates are nearly the same, the variation for circumstances being greater. It is as follows:

Feet of Surface of Boiler.	Contents of the Building in Cubic Feet.
115	18 to 30 thousand.
125	26 " 43 "
133	37 " 62 "
148	55 " 92 "

The foregoing remarks and the tables just given go to show the approximate relation between boilers and radiating surfaces. In a subsequent chapter devoted to methods of calculation pertaining to steam heating data, founded more on the scientific principles of the matter, will be given, from which a better conception of the subject may be obtained by those who desire to go more thoroughly into the matter.

CHAPTER IX.

GRATES AND CHIMNEYS.

For a house heating apparatus the grate and fire-pot should be so constructed that as the fire burns the body of fuel will move together, centrally as well as downwards, and thus keep a compact body of ignited coal for a long time on the grate. When a grate is broad, with a thin fire on it, as in power boilers, the fire burns out at certain parts of the grate faster than at others, and a fireman has to build his fire accordingly, giving it constant attention to keep up steam and not waste coal; but in a private house, all parts of the apparatus, including the grate and fire-box, must be constructed so that the fire can be left unattended for a comparatively long time; and engineers unacquainted with this class of work will be surprised at what has been done in this respect, 8 to 12 hours' duration being common for a fire to keep steam, and often make a better showing for the same weight of coal per radiating surface than large boilers with flat rectangular grates, fired regularly and often, with a high rate of combustion.

When a grate is surrounded with a fire-pot, or when the fire-box is drawn in to any angle not greater than about 30° from the perpendicular, the coal as it burns will press to the center and slip down, keeping the fire deep and in a good condition longer than when a furnace has perpendicular sides. The tapering fire-pot works well with a magazine-fed fire, as the tendency is to consolidate the fuel as it slips downward. Ordinarily, however, the sides of the fire-pot or fire-box are perpendicular, in which case it must be deep to hold fire enough for a whole night in cold weather.

Grates should be proportioned to the heating surface of the building (radiating surface), which, of course, is proportional to the water to be evaporated or the steam to be condensed.

Ordinarily, a pound of anthracite coal will evaporate ten pounds of water from the temperature of the return water to steam, at almost any pressure. To evaporate water—make steam—from water, at 178° Fahr., a fair temperature for return water, to one pound pressure of steam 1,000 heat units are required. This is low pressure. To **evaporate** the same water to 100 pounds pressure of steam it will require 1,047 heat units. This will be for high pressure conditions. If, however, we still have high pressure conditions, and a very much hotter temperature of the return water than before—say a temperature of 225° Fahr. for the return water—which would be a natural condition with a high pressure gravity return apparatus at 100 pounds pressure, then the heat to evaporate one pound of the return water to steam again, 100 pounds, will be still 1,000 heat units, or the same as for low pressure steam. So that, approximately and for all ordinary usages, it is safe enough to say that one pound of water requires 1,000 heat units to evaporate it, no matter what the pressure. With this in mind, we have the first essential item of data in finding our grate surface. Suppose we are going to evaporate 1,000 pounds of water in an hour, then we know we must burn about 100 pounds of coal in the same time —one hour—and this being fixed, we proceed to determine the size of the grate.

It has been found by many experiments on American coal, that when it is consumed at the rate of between 8 to 9 pounds per hour per square foot of grate, that the maximum of practical efficiency is obtained with ordinary grates and boilers. This seems to establish at once the grate surface that should be used, but in this we would be incorrect for all purposes. For high pressure boilers, with a fireman in attendance, it has been established, empirically, that one square foot of grate should burn 10 pounds of coal in the hour and this is generally the proportion used in common boilers, where the conditions of draft, etc., is not known. But for house heating boilers this will not do. It is not the question of whether a certain area of grate to a given weight of coal gives the greatest efficiency per pound of coal, but it is a question of preparing a grate and furnace that will hold coal sufficient for a night's burning without further attendance than to clean and fill the furnace at bedtime and find steam in the morning at 6 or 7 o'clock, without unnecessary waste of fuel. This is the condition presented in house heating, and it is for this we must proportion a grate and furnace in a heating apparatus. If we are to evaporate 300 pounds of water every hour, from 10 P. M. to 7 A. M. (9 hours), we know we must burn at least 270 lbs. of coal during the night. A quantity of anthracite coal of this weight will be from 7 to 8 cubic feet in bulk, and if we give it a depth of one foot over a grate, it will require a grate surface of from 7 to 8 square feet. A fire 1 foot thick is a deep fire, but as it is to burn so slowly, sufficient air will pass through it with any ordinarily good draught. Even if it had to be 14 or 15 inches thick, to run a couple of hours longer, this probably would not matter; so that we have established a fact that in this class of cases our grate must be, say, 8 square feet superficial area to burn 270 pounds of coal in 9 hours, or 30 pounds of coal per hour, giving us a ratio of one square foot of grate to each $3\frac{3}{4}$ pounds of coal burned per hour. Now practice has demonstrated that a rate of combustion of about 4 pounds of coal per hour per square foot of grate is a proper and reasonable consumption for house heating boilers. Less grate area may do with a good draft and thicker fires, but the chances are against efficiency in coal consumption when the 4 to 1 limit is passed, that is, when more than 4 pounds of coal per hour are burned. With this proportion of grate to coal, the accumulated ashes will not prevent the passage of the proper quantity of air as the time for firing again approaches, and it may safely be relied upon.

The amount of air space in a grate must not be overlooked. If the air space is contracted or very fine, more intensity in draft is required. This is why some apparatus does better with one style of grate than with another. The total area of the grate may be near the regulation size, but the air space is not sufficient or it is of such a nature that it becomes choked with the ashes and clinker too readily, thus the *openings* in any grate must be sufficiently large to pass the greatest quantity of air required when the fire is packed with ashes, as in the last hour it is supposed to run without attendance. Smaller openings will not answer, and any much larger are unnecessary, although there is considerable scope in this latter respect as it is the constant opening or closing of the draft-door which really regulates the quantity of air required by the fuel, provided it is ample in the first place.

CHIMNEYS.—The question of the chimney should be considered with the subject of grates. There are two requisites for all chimneys: First, a chimney must be able to pass air in sufficient quantities to consume the coal, and, second, the intensity of the draft must be equal to passing the required quantity of air through the fire, no matter how much ashes there may be on the grate nor how thick the fire may be. The size of the coal used may be changed so as to favor a low or poor intensity of the draft, but should the air supply be insufficient in quantity, intensity will do very little to make it up and the result will be insufficient steam or no steam, and a constant poking of the fire without satisfactory results. Coal will burn in an open grate or on the hearth, but the rate of combustion cannot be controlled, and the mass simply burns from the outside unless the blower is used. A chimney 40 feet high will generally have the required intensity, but this same chimney may be so reduced in volume by bad turns in the wall, or by insufficient area in any part of its length or its whole length, that it cannot burn the coal required, and therefore is useless. It sometimes happens that the chimney has just about the cross-sectional area that will do the work under favorable conditions of weather and fuel. These are the worst kind of chimneys. They pass muster under the favorable condition of affairs, and fail entirely in damp and cold weather, or with unfavorable fuel, such as the harder kinds of anthracite coal, when with the free burning coals they can be just made to work by watching them. When a chimney proves itself entirely too small, it is, of course, improved by enlargement or by an increase of height, though in the latter direction much is not to be expected from an increase of height of 10 or 15 feet, as the velocity of draft increases very slowly indeed with the increase of height of the chimney.

It is well to remark that a chimney may be too large in diameter, though this does not often occur. Still I had a 12×36 inch chimney, of about 40 feet high, into which a small boiler with about 400 feet of radiating surface connected. 1 found it necessary to divide this near the middle by a wall, leaving a 12×16 inch flue for the boiler, before satisfactory results were obtained. The amount of heated gases passing into the large chimney was not enough to cause a movement of air through its whole section, so that the intensity became almost nothing, and the conditions were not much better than passing a stovepipe through a hole in a wall into an area or light shop.

Thus the chimney must be capable of passing sufficient air for the greatest consumption of fuel ever likely to be used in the apparatus. Less air will not do. More than is needed does no harm, for it is within the power of the operator or the automatic draft regulator to diminish the quantity of the air.

An old rule is that the area of a chimney should be

not less than one-eighth the area of the grate. If this rule was correct, or nearly correct, for mill chimneys, it cannot blindly be applied to chimneys in private dwellings, though, on the other hand, if a designer has no better guide, it may be followed with some degree of surety.

But the 1 square foot of chimney area to 8 square feet of grate, as applied to high-pressure boilers, where the consumption of fuel is about 10 pounds of coal to a square foot of grate, will not do in all cases where the consumption of coal falls to as little as 4 pounds per foot of grate, or the 1 to 8 rule would result in an enormously large chimney.

Suppose the case of a hospital or school with a chimney 100 feet high, arranged on the gravity automatic principle, the intention being to keep a fire over night without attendance, in which the grate area might be from 80 to 100 square feet, the 1 to 8 rule would be a chimney out of all proportion to the size and actual requirements of the building, though in the case of a large private residence where the total grate area would be only 8 square feet, a chimney of 1 square foot, 40 to 50 feet high, would be eminently proper, as such a size would be required to give a practical magnitude to the chimney.

The proportion of frictional surface in a small chimney is very much greater than in a large chimney, and thus the 1 to 8 rule will do very well in chimneys for house work, or where the chimney seldom exceeds 40 feet in total height. In very small apparatus, however, the 1 to 8 rule may not prove ample. Take a grate of $3\frac{1}{2}$ square feet; under the 1 to 8 rule the chimney will be 8×8 inches. Now an 8×8 inch chimney is usually enough for the ordinary American stove, but short 8 × 8 inch chimneys often fail to give the required draft for a steam heating apparatus with a grate of 3 to 4 square feet. If the chimney is built smooth and straight, and 30 to 40 feet high, it may prove just ample, but if it is drawn over two or three times, to carry it around fireplaces at the different stories, and roughly cobbled on the inside with chimney pots and other apparatus on its top, the 1 to 8 rule may prove very deficient.

An 8 × 12 inch chimney is the smallest that should be built in a house for a heating apparatus, though not because it may actually require that size chimney for the combustion of the coal, but to give a practical magnitude for roughness and want of cleaning, etc., and no other pipe or flue should be taken into it except the boiler flue.

For apparatus, such as are put into large mansions, which burn 40 to 50 tons of coal in 180 days, a 12 × 16 inch flue is little enough for the above reasons.

Care in building a chimney is necessary, as a smooth chimney will give a better draft and keep clean longer than any other. Offsets in chimneys should be avoided, and equilateral and parallel sides are best unless the chimney can be round.

To those who are interested in large chimney construction I would recommend R. M. & F. J. Bancroft's practical treatise (Jno. Calvert, publisher, Manchester, Eng.), where theoretical formulas are given, together with illustrations of many large chimneys, with the results of experiments obtained therefrom.

The intensity of a chimney, due to its height and heat, is usually represented by inches of water. For instance, a chimney 100 feet high, warmed until the bulk of its gases has been doubled, or the density of the air in the chimney reduced one-half, will have an intensity or power nearly equal to three-quarters of an inch of water. Remember, however, that this intensity of three-quarters of an inch head of water can only be obtained in such a chimney at the moment the damper is shut and the flow of air checked in the chimney. When the damper is open and the air or gases passing freely through the chimney, another condition of intensity exists much less than the former, and which may fall to a third or quarter of it, or even less, only one-tenth, according as the chimney is passing less or more air. The theoretical intensity in a chimney 100 feet high, in which the gases have been doubled in bulk by raising their temperature 500° Fahr., will give a theoretical velocity of about 112 feet per second in the middle of the chimney, but as this 112 feet per second is dependent on the intensity, and as the intensity begins to decrease the moment the air begins to move, a new condition follows, the result of intensity becoming less and quantity greater, which may be called the practical efficiency of the chimney.

This practical efficiency constantly changes in chimneys, depending on their height, temperature, area and shape.

It is of more importance to us, therefore, to be able to discover the practical efficiency of a chimney, or, more properly, to construct a chimney that will give us a desired efficiency, than to find the theoretical velocity or intensity, as they both vanish in practice and resolve themselves into the third condition, which is the one we are interested in. The friction of the chimney is the largest and most important factor in producing this third condition, although the weight of the gases, compared to air, that pass through the chimney have also something to do with it. The theoretical velocity must form the basis of the theoretical quantity, but this is figured on the difference in weight between two columns of pure air without moisture, forgetting there is carbonic acid gas and unconsumed carbon in one of the columns. Now our chimney flue in practice is filled with a composition of air, carbon, etc., that has a greater specific gravity than air. Pure air is composed principally of 77 per cent. of nitrogen and 23 per cent. oxygen (by weight), and at a temperature of 60° a bulk equal to 14 cubic feet will weigh a pound under the ordinary conditions of our atmosphere. About 11 to 12 times this quantity of air is necessary for the theoretical combustion of one pound of coal, though in practice it runs up to 17 or 18 times that quantity, and sometimes as high as 20 times, with poorly brick-set boilers.

The practical efficiency of a chimney may be measured by a water-gauge when the apparatus is in full operation, with the dampers wide open while passing through the fuel the full quantity of air required for proper combustion, so that the gases of combustion, instead of being of the same weight as common air, will be about one-twelfth heavier, reducing the theoretical efficiency considerably. The friction in the chimney, the friction of the smoke pipe and its turns, the friction of the boiler and its flues, and the friction of the air through the grate and its bed of fuel, all tend to cut down the theoretical velocity and intensity so much that a very high factor for safety must be provided in a chimney constructed by theoretical rules. Still, without theory much could not be done. It establishes safe comparisons and an ultimate limit beyond which it is impossible to go without using mechanical power to increase the velocity of draft.

Presumably, if any rough rule was advanced for chimneys, it would not be safe to assume a greater intensity than one-tenth the theoretical value, and for small chimneys that would probably be too great. In a chimney 100 feet high, warmed about 500° above the outer air, the intensity is about 0.75 of an inch of water, and the velocity corresponding thereto, about 112 feet per second in the chimney, and about half that, or 56 feet per second through the ash door, if equal area with the chimney, as the air is about double the density and half the volume before it passes through the fire. Now, for an intensity of one-tenth, or 0.075 of an inch of water, the flow of gases in the chimney will be about 38 feet per second, and at the draft door about 19 feet per second, or a practical efficiency of about one-third. These are high practical efficiencies and are not to be obtained in small crooked chimneys built into the walls of a house and often into wet or damp outside walls.

The theoretical intensity of a chimney may be obtained and remembered in a practical way as follows: One pound of pure air fills 12.387 cubic feet at a temperature of 32° Fahr. This will make a column of air of one square inch 1,783 feet high, which, of course, will exert a pressure at the foot of the column of one pound per square inch. We never, however, have to do with so great a pressure in an ordinary chimney draft as one pound, but we have to work with ounces or inches of water. The ounce of pressure will sustain a column of air $111\frac{4}{10}$ths feet high (which for easy remembrance may be called 112 feet), and an inch of water pressure will sustain a column of air 66 feet high or two-thirds of 100. feet, which is also easy of remembrance. If we increase the temperature of the column of air 490° Fahr. we double its height and make its density only $\frac{1}{2}$, or, if instead of increasing its temperature 490° Fahr. we increase it only 245° Fahr., we increase its height one-half, or 33 feet, making the equivalent height of the column 100 feet. Now this is an ordinary condition for the chimney of a low pressure steam apparatus. The gases are at a temperature of from 250° to 300° Fahr. as they enter the chimney, and the density of the column of air, though it is 100 feet long, is only equal to the density of a column a little over 66 feet long at the outside temperature. Now it is the difference in height between these two columns that gives the data for the velocity. This difference, called *head*, or difference of height, is 33 feet, and the velocity of the air as it enters the flue, due to such a height, will be the same as a stone would have in that part of its travel when it reached 33 from the point it first fell from. This being known, the remainder of the calculation is easy, it being simply to find the velocity of the stone by the rule $V = \sqrt{H} \times 8$, in which V is the velocity looked for and H the height in feet fallen through, and 8 half the distance fallen through in a second of time. Our height being 33 feet, the square root is 5.75 feet, which multiplied by 8 gives 46.92 feet as the theoretical velocity per second. The column being warmed and expanded one-half its original length, the intensity will be one-half the original weight.

If we have a chimney still 100 feet high and we warm the gases until they are double their bulk, then the density is $\frac{1}{2}$, and the head or height one-half the chimney. The chimney being 100 feet, H is 50 feet, square root of which is 7.07, which multiplied by 8 gives 49.56 as the theoretical velocity—not a very large gain for an increase of about 245° Fahr. in the temperature of the chimney gases.

In practice I do not dare to give the ordinary 100 feet stack a greater efficiency than that due to a velocity of 25 feet per second, and when the walls of the building will admit of good nearly equilateral chimneys I figure it on the basis of 15 feet per second, with the bulk of the gases taken at 600 cubic feet per pound of coal.

As the subject of grates and chimneys is treated together I will again refer to grates before finishing the chapter, as the same order was carried out in the earlier editions of the book.

Diameter of Round Grates in Inches.	Square Feet of Surface in Grate.	Diameter of Round Grates in Inches.	Square Feet of Surface in Grate.
$13\frac{1}{4}$ inches.	1 feet	$26\frac{1}{4}$ inches.	$3\frac{3}{4}$ feet.
15 "	$1\frac{1}{4}$ "	$27\frac{7}{10}$ "	4 "
$16\frac{1}{4}$ "	$1\frac{1}{2}$ "	28 "	$4\frac{1}{4}$ "
18 "	$1\frac{3}{4}$ "	$28\frac{3}{4}$ "	$4\frac{1}{2}$ "
$19\frac{2}{10}$ "	2 "	$29\frac{1}{2}$ "	$4\frac{3}{4}$ "
$20\frac{1}{4}$ "	$2\frac{1}{4}$ "	$30\frac{1}{4}$ "	5 "
$21\frac{1}{4}$ "	$2\frac{1}{2}$ "	31 "	$5\frac{1}{4}$ "
$22\frac{1}{4}$ "	$2\frac{3}{4}$ "	$31\frac{3}{4}$ "	$5\frac{1}{2}$ "
$23\frac{1}{2}$ "	3 "	$32\frac{1}{4}$ "	$5\frac{3}{4}$ "
$24\frac{1}{2}$ "	$3\frac{1}{4}$ "	$33\frac{2}{10}$ "	6 "
$25\frac{1}{4}$ "	$3\frac{1}{2}$ "		

112   STEAM HEATING FOR BUILDINGS.

The preceding table gives the number of inches in diameter for circular grates, from one square foot to six, inclusive, advancing by one-quarter of a square foot, and will do for ready reference by the fitter.

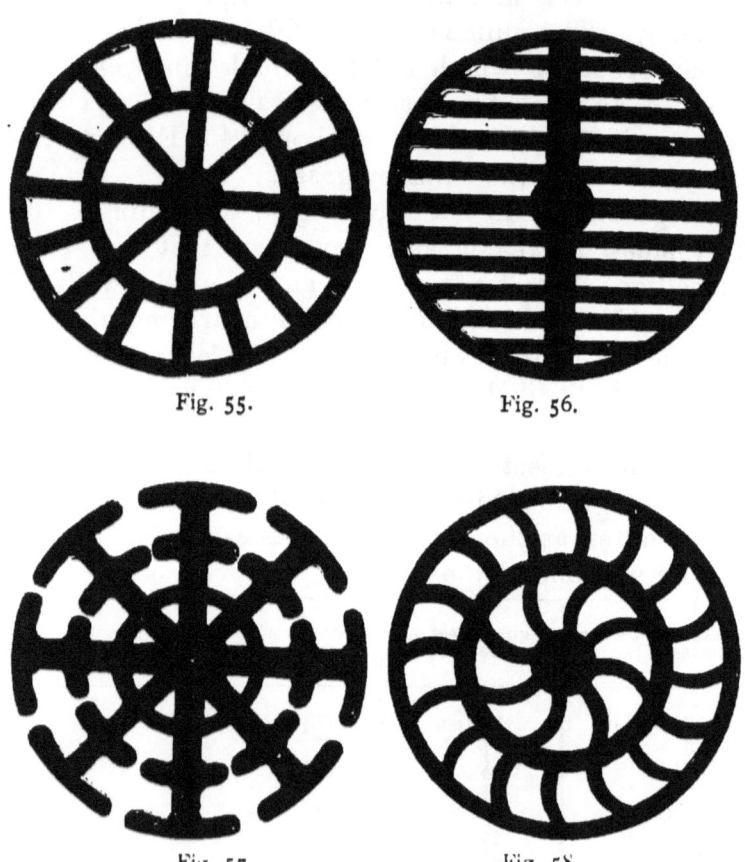

Fig. 55.   Fig. 56.

Fig. 57.   Fig. 58.

Why do grates break? Round grates made of concentric rings and straight radial arms always

break and fall to pieces, never wearing out in the ordinary way. There is usually the same result with parallel bars, confined with a ring, and they are the two forms most likely to be made by any one who is required to get up patterns and has not had experience in the matter; since the pattern for the straight-barred grate is so much easier to make. The reason for its breaking is because the thrust of the straight parts of the grate is not compensated for when expansion takes place, and a rupture of the outer rings is the result.

In this matter it would be well for the engineer to take pattern from the stove manufacturer, and follow him in this respect. No straight bars are here used in circular grates, as a rule; or, if one has to use straight bars, they are short and unconfined at one end, radiating in or out.

The same principle applies to *all* grates. The old-fashioned three-barred common grate fails by reason of the ends dropping off when least expected, due to the unequal thrust of the bars against their ends, quietly cracking them in the angles, where they are the weakest. Figs. 55 and 56 show grates that will crack; Figs. 57 and 58 show grates which will not crack, if very sharp corner angles are avoided by rounding them a little.

Shaking grates have taken the place of common grates in nearly all styles of heating apparatus, and in the latter part of the book the subject of grates will be taken up again and gone into more in detail.

CHAPTER X.

SAFETY-VALVES.

Every boiler, for the generation of steam, for power or heating, *must* have a safety-valve.

A perfect safety-valve is a desideratum, for with a valve of sufficient area that will respond to the desired pressure of steam an explosion from *over-pressure* would be an impossibility.

The primary use of a safety-valve on the steam generator is to relieve an excess of pressure, but, aside from that, the noise that it produces when the steam is escaping makes this an auxiliary to the pressure-gauge by indicating that the maximum pressure has been reached, and that immediate attention of the engineer is required, if not in the interest of safety, at least in the interest of economy.

A safety-valve, to be sufficiently large, should be of such proportions that it will let all the steam escape which the strongest fire is capable of producing when all the other outlets of the boiler are closed, and for house boilers particularly the safety-valve should be calculated by a rule based on the greatest evaporation.

Engineers who have given sizes for power boilers have not considered this question in relation to house boilers, and nearly all have a different rule for the questions they have considered, with frequently very different results.

To a steam-fitter looking for information on the subject, and who is confessedly ignorant of the problem, this diversity must raise doubts as to the authenticity of some of the methods, and he is liable to be guided by the general reputation of the writer as an engineer, and take it for granted that his rule is applicable to all cases, and apply it to house-heating purposes.

All boilers should have ample safety-valves, but house boilers which are automatically governed and, as is customary, left for long periods without any one near them, the safety-valves will be the sole regulators, should their regulating doors fail, and consequently they must have proportions beyond a doubt as to their efficiency.

Many boilers burst when working at their *ordinary pressure*, and mysterious unavoidable causes are often assigned as the reason; but there is only one reason—*insufficient strength*, and that either from a defect of construction, or by deterioration of the material, or burning through neglect; and in a case of this kind no safety-valve can respond, the valve being set for a higher pressure than that at which the boiler explodes.*

---

* I have entered boilers where pins were out of braces, and braces broken; and one case where the mud deposit in a horizontal boiler covered four rows of tubes at the back end, cracking and bulging the shell, the bank of mud apparently holding the boiler together.

The office of the safety-valve being to relieve the boiler of pressure above its ordinary working pressure, it must be large enough to let the greatest quantity of steam ever likely to be made escape freely.

In proportioning safety-valves for small boilers, and, in fact, for most boilers, the size is frequently simply guessed at; the engineer or fitter puts on a valve of certain size, because he is in the habit of doing so, or because some former employer did it, having in mind the while an idea that if a certain size pipe carried all the steam the boiler could make to the engine, a safety-valve very much smaller in area would answer, since it escaped into the atmosphere *only*—not knowing that a two-inch safety-valve blowing off at 60 pounds had an opening so small that if it was round he could not put his pencil through it.

When a valve begins to blow off, the pressure underneath the center of disk decreases out of all conceptional proportions to the pressure in the boiler; the decrease not being due to a diminution of the pressure in the boiler (as the steam may actually be increasing), but to the draught caused by the escape. The laws of the phenomenon are imperfectly understood, but the results have been conclusively confirmed by Professor Trowbridge and others; the proportional difference being greater for greater pressures.

Professor Burg, of Vienna, found by actual experiments with an apparatus constructed for the purpose, that a valve of 4 inches diameter raised from its seat when blowing off, according to the *two* first columns of the following table. The last two columns are calculated that the fitter may form an actual conception of the openings by comparing them to something he is perfectly acquainted with.

The first column gives the pounds per square inch; the second, the actual lift in fractions of an inch; the third, the actual size of openings in decimals of a square inch, when the bevel of the valve seat is 45 degrees; and the fourth, the internal nominal size of gas pipe nearest the actual opening.

TABLE No. 2.

1. Press.	2. Lift.	3. Area.	4. Pipe.
12	$\frac{1}{36}$	.25	$\frac{1}{2}$
20	$\frac{1}{48}$	.187	$\frac{3}{8}$
35	$\frac{1}{64}$	.166	
45	$\frac{1}{68}$	.137	
50	$\frac{1}{86}$	.1043	$\frac{1}{4}$
90	$\frac{1}{168}$	.0534	$\frac{1}{8}$

The following graphic illustration has been made to show at a glance the size of the openings:

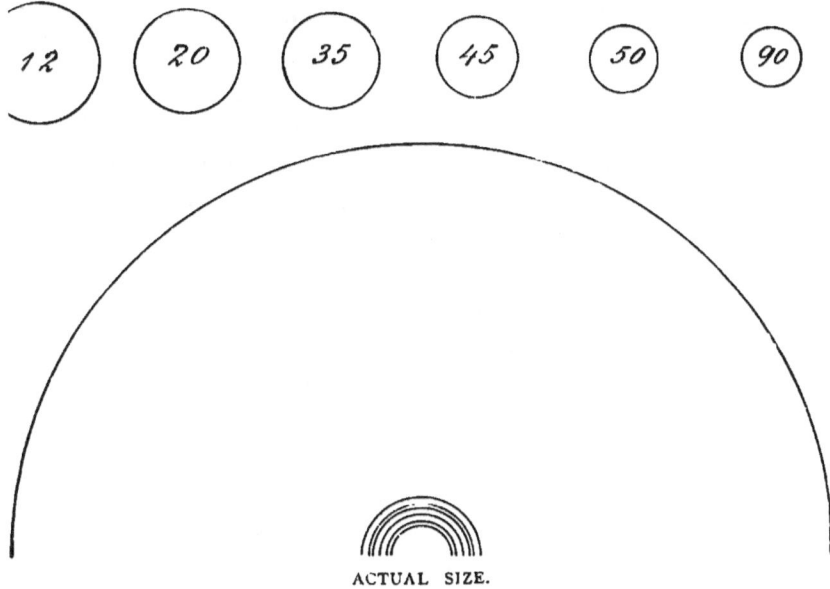

ACTUAL SIZE.

The large rim incloses the area of a 4-inch disk (12.56 square inches), and the smaller ones the areas of the openings at the different pressures.

It can be seen from the foregoing that an increase of pressure lessens the size of the opening; nor do the increased pressure and flow of the steam compensate for the decrease in the size of the opening, and what is required is a valve of very great diameter, or one that will open nearly its full area.

There are many formulas for calculating the size of safety valves, all based on the size of the disk; and, though arbitrary, they may be useful, as they give sizes about *four* times the area of ordinary practice.

Fairbairn allows 29 square inches for a 50-horsepower boiler.

This valve is nearest to what can be purchased as a six-inch valve. When we take into consideration the variations in formulas for velocity of flow of steam through apertures, that these velocities are based on the efflux through round holes or conoides, that the lift of the circular six-inch disk must be sufficiently great to allow an annular opening equal to $\frac{3}{4}$ of a square inch, and perhaps as great as $1\frac{1}{8}$ square inches for a pressure of 50 pounds above atmosphere, should the conclusions be correct which we draw from the experiments of Prof. Burg with a four-inch valve (vide Trowbridge, " Heat Engines "), we are not sure that it is possible to get an opening equal to the smallest above mentioned.

Rankine says: " Divide the number of pounds of water which enters the boiler in an hour (to supply the loss by evaporation) by 150, and the product is the area of the valve in inches."

Bourne says: "Multiply the area of the piston in inches by its velocity in feet per minute, and divide by 300 times the pressure of the steam, and the product is the area of the valve in inches."

Another rule allows one-half a square inch of the disk for each square foot of grate, and though I am of the opinion that the opening in valves should be proportional to the grate surface, the question will arise why is not pressure taken into consideration, even assuming the half-inch of the valve area to be sufficient. But is it enough? With ordinary boilers it is common to burn 10 pounds of coal per hour on a square foot of grate, and should the coal evaporate ten times its own weight of water (not an impossible thing) there will be 100 pounds of water evaporated to a square foot of grate in an hour, which will give a 50 H.-P. boiler about 30 square feet of grate and 15 inches of area of valve. According to Zeuner, it will take 3000 pounds of steam about an hour and a quarter to pass through a hole with a section of one square inch when the pressure is 50 pounds. No four and one-half inch valve of common construction (the nearest to 15 square inches of disk) can give such an opening, and the more I consider it the more I am inclined to repeat the words that "the ordinary safety-valve is only a danger signal."

According to an act of Congress, for steam-boats, etc., boilers with stayed furnaces are to have 30 square inches of disk area for every 500 feet of effective heating surface, and for cylindrical boilers 24 inches of disk for the same surface. The word effective is here taken advantage of, and as a rule (by what authority it is difficult to understand) that about two-thirds only of all the surfaces of marine boilers is considered effective. Thus, if we add 250 to the 500 mentioned, we again have what might be called a 50 horse-power boiler, with 30 inches of disk or a little more than a six-inch valve. The law also regulates the size of "lock-up" valves at two inches diameter of disk for 700 feet of effective surface or less; three inches for 1,500 feet, four inches for 2,000 feet, five inches for 2,500 feet, and six inches for any boiler longer. Such valves, of course, are not expected to do much more than sound an alarm.

Rules which are based on the work done in an engine cannot be applied to boilers in apartment houses and stores, in which, though the latter have engines frequently, and both have large pumps for hydraulic elevators, the steam used in the cylinders seldom represents more than one-third of the water evaporated, the remainder being for warming, cooking, washing, drying and other purposes. Again, rules based on evaporation, which do not take pressure into consideration, must be carefully sifted, and none used that will not provide for the taking away of all the steam at all ranges of pressure. Should a valve be found upon experiment to be just sufficient to relieve a boiler at 100 pounds of steam, the same valve would not do for ten pounds maximum pressure.

According to the *relative volume* of steam, at *half its theoretical velocity* when flowing into the air, two square inches of actual opening of valve should be ample for the number of cubic feet of water evaporated per minute at the different pressures given in the following table:

Pressure in Boiler above Atmosphere.	Cubic Feet Water Evaporated per Minute.	Actual Size of Opening in Valve.
1	.25	2 square inches
25	.80	2 " "
50	1.25	2 " "
100	2.13	2 " "

The ¼ cubic foot of water per minute is equivalent to about a 30 horse-power boiler, and the others are respectively 100 horse-power, 150 horse-power, and 250 horse-power; and two square inches is the smallest safe area of opening that will keep the steam down to the pressure in the first column. Of course, if we can get the theoretical velocity of flow, about one square inch area will do; but there is no factor for safety to cover friction under the valve, in the escape pipes, etc.

By a study of the above, it will be seen that if a boiler is of such construction that 25 pounds of steam is the maximum, it will require a larger valve for the same amount of water evaporated than a high pressure boiler, and that indiscriminate rules are not to be used.

There has been much effort to obtain a safety valve which will give a large opening, and in some instances valves thus made have proved practically a success, though not in general use, since the necessity for them is not recognized by the public, who content themselves with a *danger signal*, where the noise it makes when blowing off, is all that can entitle it to the name of safety valve.

Fig. 59 shows a common safety valve, with an auxiliary attachment, which is capable of pulling the valve open to its full extent. *A* is an ordinary safety valve, put on in the regular way; *B*, a common low-pressure diaphragm or regulator, to be described later, attached to the end of the lever, and suitably fastened to the boiler with the pipe connection *C*, to the under side of the diaphragm, and taken from the water space of the boiler, for two reasons,—namely, that the water in the pipe may be cold, so as not to affect the rubber of the diaphragm. The water being steady and solid prevents vibrations, and gives the initial pressure unaffected to the underside of the rubber. Fig. 60 shows the same apparatus in a position when blowing off, the pressure under the rubber overcoming the weight on the lever.

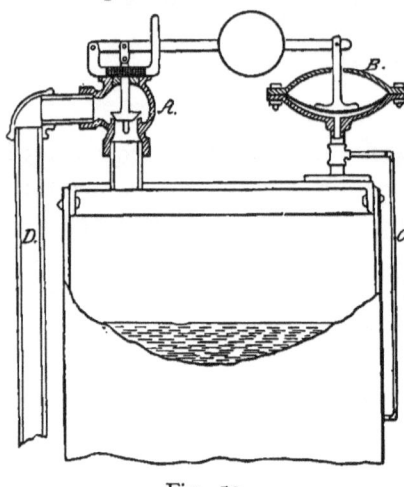

Fig. 59.

When steam begins to escape, it cannot affect the diaphragm until the pressure in the boiler falls, when the diaphragm subsides.

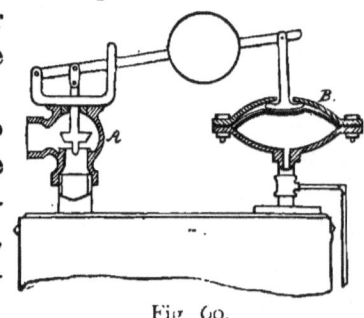

Fig. 60.

This same principle can be applied to high pressure safety valves by using a diaphragm, especially constructed, as in high pressure damper regulators.

The escape pipe *D*, Fig. 59, of the safety valve, is sometimes carried down and under the grate by steam-fitters, in order that the escaping steam may dampen the fire, and check it by interfering with

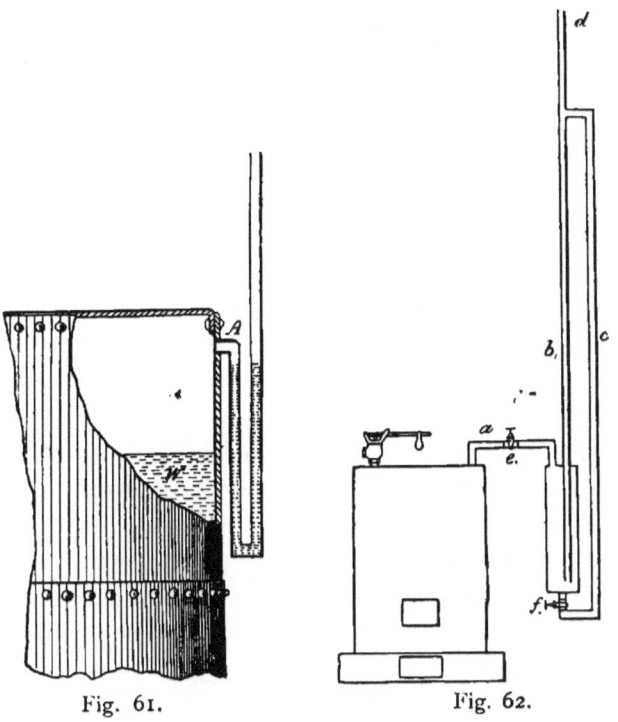

Fig. 61.   Fig. 62.

combustion, a point worthy of consideration by all engineers.

Another arrangement for very low pressure is a water column, connected as in Fig. 61. A connection, *A*, is taken from the steam space, and carried down and up, forming an inverted siphon filled with water.

When the pressure in the boiler exceeds the weight of the column of water in the pipe, it blows it out, letting the steam escape, which will blow until the steam is all gone, or the pipe again filled with water.

A modification of this principle has been constructed, by which steam can be carried to about 12 pounds per square inch, in buildings of ordinary height. A cylinder of any suitable construction is connected to the boiler, as shown in Fig. 62, and filled with water; the pressure of the steam through the pipe a on the surface of the water in the cylinder presses it up in the pipe b; but when the pressure is great enough to send the water over into the pipe c, the steam escapes at d. This arrangement, like the one before, will not stop blowing without manipulation, it being necessary to close the valve e, and open the valve f, to let the water again into the cylinder.

A boiler with this arrangement on it, should also have a common safety valve set at a lower pressure, to give warning, for should this start to blow off, and be neglected, it will waste water and steam from the boiler. The pipe a may be long, so as to have the cylinder a considerable distance from the boiler; in one case where it was set against it, the heat evaporated the water from the cylinder.

A boiler with a safety water column on it, as described, should have a vacuum valve also, to prevent the water from being drawn into the boiler when steam goes down.

Another arrangement which has been tried with some success, is an ordinary safety valve of large size, with a pipe a carried from the under side of the disk down into the water in the boiler, as shown in Fig. 63; the orifice of the valve forming an annular space around the pipe.

The principle of this valve is that the pipe being carried down into the water represents a certain area of the disk, which should be of scarcely any value

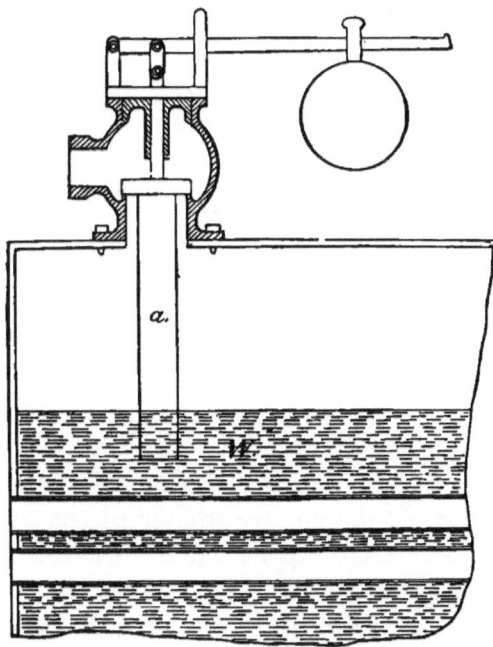

Fig. 63.

when blowing off, but by being in the water the pressure underneath is not relieved.

Pop-safety valves with differential disks and seats are also used for high pressures, by which very much larger steam passages can be secured than with ordinary valves of the same diameter. Details of some safety valves will be given in one of the latter chapters of the book. There are pop-safety valves on the market that will open and blow down a boiler in a few minutes, not only letting the excess of steam escape, but capable of allowing all the steam the boiler is capable of making, escape freely.

Another point of interest to a fitter is an easy method of finding the weight necessary for a safety valve.

If they reason as follows they will always be able to find the required weight within practical limits and approximately correct. If the area of the valve disk is one square inch, it is only necessary to have a one-pound weight placed on the top of it (assuming itself, the disk, to have no weight) to keep it closed against one pound per square inch of steam, and, of course, if the weight is 10 pounds it will hold the disk against 10 pounds per square inch. This is substantially the "low pressure safety valve" with the weight on a spindle on the back of the disk; the weight of the disk and of the "weight" having a total effect (without leverage, of the square inches of the valve multiplied by the maximum pressure to be carried. But with valves having levers for high pressures, wherein the weight must be fixed and kept within ordinary limits of size, the relative position of the weight to the disk must be considered, and the weight of the lever also.

The lever always represents a constant resistance to the valve, and should be found first. For instance, with a disk of one square inch, with a lever of one pound, the lever being ten inches long, as shown, and the distance from the valve spindle to the fulcrum one inch, the lever will always exert the same pressure on the disk as a one-pound weight would, placed on the lever at half its length (five inches). Thus it may be borne in mind that the lever is a weight hung on an imaginary lever of half its own length and exerting pressure according to its weight.

In this case, where the lever is one pound, the distance from the fulcrum to the spindle one inch, and the whole length ten inches, the lever will exert five pounds on the valve without a weight. But it is necessary to have a weight which is movable, as in Fig. 64, and suppose we have one of one pound also. If we placed this weight now on top of the spindle it adds just one pound to the force exerted by the lever $(5 + 1 = 6)$, making the total pressure at that point

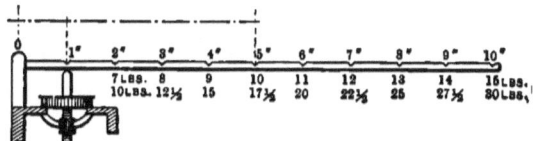

Fig. 64.

six pounds; but if we move it the same distance from the spindle that the fulcrum is, only in the opposite direction (out on the lever), it will exert two pounds additional to the pressure exerted by the lever and will exert an additional pound per inch as it is moved from the disk, giving for the end of the lever 15 pounds and for the second notch seven.

If, instead of a one pound ball or weight we have a $2\frac{1}{2}$ pound ball, the pressure exerted by it will be the weight, $2\frac{1}{2}$ pounds, multiplied by the distance to the first notch (2 inches), equal to 5 pounds plus 5 pounds for the lever, making 10 pounds, or with the weight at the last notch, $2\frac{1}{2} \times 10 (= 25) + 5 = 30$ pounds.

From this we get the simple formula $A \times P \times D \div W = B$, in which—

A is the area of the valve disk in square inches;

P is the pressure of steam in pounds per square inches;

D is the distance between the stem of the valve and the fulcrum;

W is the weight of the disk, stem and lever;

B is the weight of the ball in pounds.

CHAPTER XI.

DRAFT REGULATORS.

WHEN the steam-heater wishes to govern anything automatically, his first thought is whether a diaphragm will answer, and if he can regulate what he wants with a rubber or light metal diaphragm, he will never resort to a moveable piston, knowing the diaphragm will work until it wears out without getting out of order, and that a piston must be kept in the nicest of order to be depended on, since it is affected by corrosion and dust, while the diaphragm, being simple and cheap in construction, and having no delicate parts, will respond to small differences of pressure and will run for many years when constructed and put on by one who understands it.*

The steam-fitter uses it to regulate the ash-pit door, for the admission of the proper quantity of air to the fire in order to govern the steam pressure; to open the fire-door so that cold air is admitted through the furnace in case the draft-door is neglected (by leaving a clinker or lump of coal underneath the edge); to open the safety-valve, and sometimes to open a "break

---

* For high pressure, nicely fitted pistons have given excellent satisfaction for damper regulation. They will not do for very low pressures however.

draft," an opening in the chimney. He also uses it for regulating the air supply to indirect radiators, to govern the pressure of steam when expanding from high to low pressures in different systems, and to regulate water pressures.

Fig. 65 shows a regulator of ordinary construction, with a bowl at the top and bottom of the diaphragm, in which A is the bottom bowl, to which the

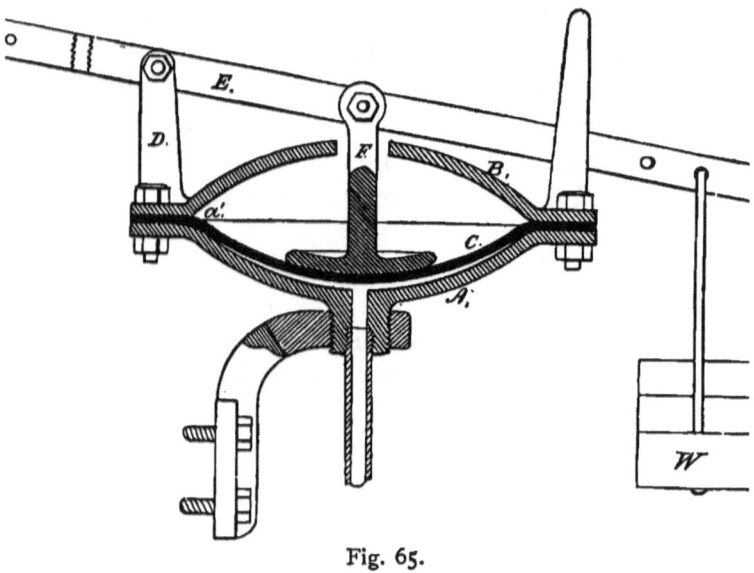

Fig. 65.

support and pipe are attached ; B, the upper bowl, to which the fulcrum and lever are attached ; C, the diaphragm ; D, the fulcrum ; E, the lever ; and W, the weight; the pressure under the diaphragm being the operating force.

In constructing regulators, sharp edges of the metal should not be left to cut the rubber. The corners of the bowl at a' should be nicely rounded, and the flanges around the edge should be deep, to give room for the bolt holes, so that they will not be too near the inner edge. The standard F should not be riveted to the rubber, but rounded on the bottom to lay on it; nor should there be holes made in the rubber for any purpose inside of the holes in the flanges.

Common flat rubber does not make a good diaphragm; it should be of extra good quality, thick, and dished to fit the bowls; so that when inflated, there will be no tension on the rubber.

Some makers leave off the upper bowl, using only a flange, but better practice requires the use of one, as it is nearly impossible then for over-pressure to burst the rubber when supported by the iron over its whole extent.

In the construction of a diaphragm for high pressure, which will not burst, it is necessary that a very small portion of the surface of the rubber should be unsupported at any time; and the movement should be small, requiring the use of a compound lever with an ordinary weight.

Fig. 66 shows a *high pressure draft regulator*, with a compound lever, in which a very small movement of the disk A will give a movement of 6 inches or so at the end of the lever at B, without straining the rubber in the least, the slackness at C forming a concentric corrugation, which admits of all the movement necessary.

In connecting diaphragms with the boiler, it is best to take the pipe from the water space, as shown in Fig. 59, at C. But when that cannot be done, it may be taken from the boiler dome or any other convenient place, except tapping into a pipe, which already has a "draft" on it (rapid passage of steam through it) for in order to prevent irregularities of pressure it is necessary to have the initial pressure constantly under the rubber.

When it is necessary to take a steam pipe to a diaphragm, instead of a water-pipe, the pipe must be *trapped* in such a manner that it will fill with water, and the capacity of the trap must be greater than the bowls of the diaphragm; so the water that has filled the trap and cooled therein, when it is pressed forward, will be sufficient to more than fill the bowls, thus always insuring cold water on the rubber.

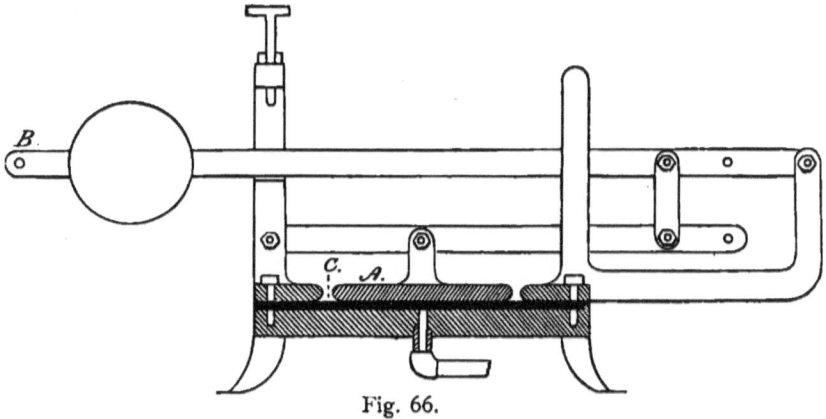

Fig. 66.

Some will not put a valve in a diaphragm pipe in a private house, fearing it may be shut off by some meddler; but this is a matter which must be left to the judgment of the fitter. A very good way is to use not less than a ⅜ pipe, and immediately under the regulator plug the pipe with iron, and bore a ⅛-inch hole through the plug. This hole will pass the wate rapidly enough for the regulator, and in case the rubber should burst, the flow of hot water will not be large.

When the rubber is fitted into the bowls without

tension, it very seldom gets holes in it, and will give warning by leaking, but should it be tight it will give away suddenly.

When regulators are attached to ash-pit doors, or to extra draft-doors, set in one side of the ash-pit (leaving the door-proper for the removal of the ashes only), the chain is fastened to the end of the lever marked G, Fig. 65, and to the door; care being taken in placing the regulator so that the chain will have a direct pull, and not interfere with the opening of other doors. When a regulator is attached to the fire-door, the other end of the lever should be used, and the regulator set a pound or so stronger than the draft-door regulator.

It is not a good plan to make one regulator do both duties, by using each end of the lever, as the doors work too close together, and a waste of fuel is the result, by letting cold air through the furnace frequently; the intention being not to open the fire-door, unless as a last resort.

85. Doors for regulators should be set at an angle of between 30 and 45 degrees from the perpendicular. When a door hangs perpendicularly with the hinges on the top (usual in such doors), the leverage changes as the door swings from the perpendicular, throwing a rapidly increasing weight on the diaphragm chain; but when the door is on a good angle the increase is not so rapid, and the door is positive in its action when closing, being hung further from its center of gravity.

Doors should be *planed* to fit tightly, and hinges and edges should be so constructed that ashes will not lodge on or under them, so as to hold them open or prevent their free action in all directions.

CHAPTER XII.

AUTOMATIC WATER FEEDERS.

The water feeders that are attached to low-pressure heating boilers, are simply regulators,—they have no power in themselves to force water into a boiler, and must be used in connection with water-

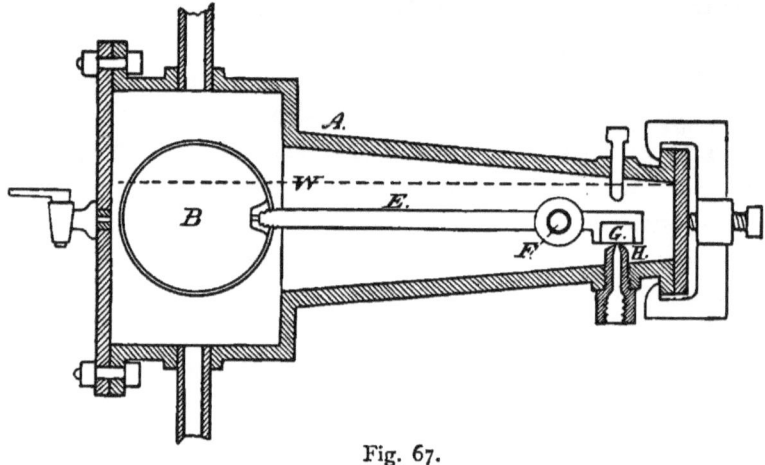

Fig. 67.

works, or a tank near the top of the house; the head of water supplying the requisite power.

So far, there has been but one description of automatic water feeder used in connection with steam-heating, and though different makers modify the shape and the valve, the principle is the same. Fig. 67 is a very good representation, in which *A* is a cast-iron case of suitable design; *B*, a copper float, with buoyancy enough for the work, and sufficiently thick so that it will not collapse with the pressure; *E*, a lever

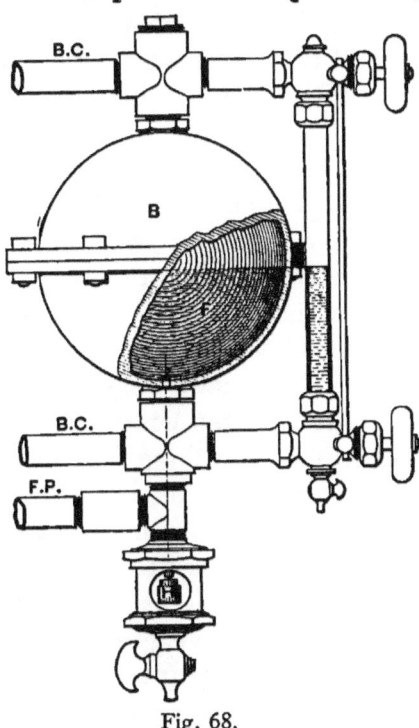

Fig. 68.

made of brass, to admit of bending; *F*, a fulcrum, and *G*, a valve, formed with a piece of hard rubber, inserted in the end of the lever, in connection with the nozzle *H*, which is usually of brass.

Fig. 68 shows a modified form of water feeder lately brought into use, in which the float acts directly on the valve, and in which the valve is visible through the glass *H*. This is very desirable, as it allows the operator to observe the valve and feed-water when it enters, and enables him to detect either a leakage or a stoppage. With this valve the pressure of the water has a tendency to close the valve, whereas with that illustrated in Fig. 67 the tendency of the pressure is to open it and cause leakage. The pipes *B C* are the boiler connections and *F P* is the feed pipe.

Copper floats in boilers under high pressure nearly always collapse ; but for low pressure they have been constructed to stand very well, though occasionally they fill with water when not well made.

Hollow copper ball floats are usually made of two pieces of copper hammered into hemispheres, and brazed together. If they could be hammered after brazing, they could be made very strong, but as the reverse is the case, and the heating to redness makes them very soft, there is nothing for the artificer to do but make them as thick as he can, without impairing their floating power too much. In the brazing of a ball together, it is necessary to leave a vent hole in one hemisphere, until the joint is thoroughly brazed, and then plug it up. A very good way to make floats for regulators, since they require some kind of a boss to fasten the lever to, is to put a boss on the inside of the hemisphere, as shown in Fig. 67, and bore a small hole through it, having the thread for the lever tapped tapering ; this hole will answer for a vent while brazing, and when ready to be fastened to the lever, the thread in the boss and the thread on the end of the lever can be tinned with soft solder and screwed together cold, which will make a perfectly water-tight joint and not leave a partial vacuum in the ball, as would happen if the ball was closed in the fire. This vacuum forms a factor not generally taken into consideration, which will materially add to the pressure the float is subject to in a boiler.

There is one point in the construction of water feeders which requires particular attention,—namely, the size of the hole in the nozzle H, Fig. 67, which forms the valve. This hole should be small, and the higher the pressure of the water-works, the smaller should be the hole. It will be seen by looking at the figure, that by the area of the hole in H, the total pressure of the water can be made to overcome the force exerted by the float. A $\frac{1}{8}$-inch hole is usually sufficient to admit all the water required; but if a larger hole is wanted, care should be taken that the ball has a preponderance; otherwise the valve will not set firmly to its seat, and the leakage will fill the boiler and prove a source of annoyance. This should be guarded against, for though it is not dangerous, it is disagreeable, and many fitters prefer to leave the feeder off on that account, since a straw, or the least dirt, will make it inoperative, and flood the boiler in consequence. In fact, the practice of to-day is to omit the automatic water feeder. When a boiler is sufficiently large to hold a quantity of water above its safe-water-line that will be equal to the amount of steam you require for the radiators of the building, the water feeder is not necessary with a gravity apparatus wherein all the water is returned to the boiler.

When there are steam-traps to any part of the apparatus, which do not return all the water directly into the boiler, the water feeder should be put on, unless there is some one constantly in attendance to supply water by some other means. With a return gravity apparatus, it may, however, be dispensed with, for the operator by looking at the water once a day, and letting in a supply when necessary, is a better reliance. A positive "open and shut" feeder, under all circumstances, has yet to be invented.

When a water feeder is used, the upper or steam pipe must not be taken as a branch from another pipe, such as the main steam pipe; it must be taken from the top of the boiler, steam header or dome, and away from other large pipes.

Special attention should be paid to the foregoing. A case which came under my notice was that of a large horizontal boiler with a water feeder connected to the dome, the water pipe entering the regular feed pipe. The feeder had a glass on it, similar to the water glass on the front of boilers, and this boiler also was furnished with an extra water glass, connected with the front tube sheet, in the ordinary way, the upper pipe being taken from very near the flange It was noticed that the water in the feeder glass always stood about five inches higher than the water in the boiler glass, which led to an investigation ; and it appeared that the water in the front glass was the true level. The upper pipe of the feeder was then taken from the dome, and tapped into the boiler shell, when both glasses showed the same level of water.

This question of draft in pipes is of vast importance, and should receive more consideration than is usually paid to it, in connection with boiler appurtenances however.

The "dancing" or fluctuations of the water in a guaged glass is sometimes caused by depression or "water-trap" in the upper pipe connection, or by the formation of steam in the lower connection. This action must not be confounded with the effect of draft. The effect of draft is to lessen the pressure on the surface of the water so that the latter will show a deceptive level. Water will never go in the glass below its level in the boiler, except momentarily, caused by a pulsation; whereas there is great danger of the glass showing a level constantly higher than in the boiler; when a water column with long connections is used, that is now so much in vogue.

CHAPTER XIII.

AIR VALVES ON RADIATORS.

THE usual position for an air valve on a coil or radiator is near the return pipe. On a vertical pipe or loop radiator it is nearly always on the last pipe or loop, meaning by the last, the part opposite the steam inlet, though in radiators with steam and return connections on the same end it is sometimes found near that end, though, in my judgment, it is better to have it opposite and furthest from the inlet, no matter where the outlet may be.

With high pressure steam the position of the air valve is not of as much importance as with low pressure or exhaust steam, and as a radiator that will work with low pressure, will *always* work with high pressure steam, it is always best to provide for the low pressure conditions.

In vertical tube radiators the valve is generally placed high up on one of the pipes, the lower end of which was sometimes run down within the base of the heater, to very near the bottom. This was done on the assumption that the air being heavier than steam, would be the first to go out by the air-vent, and is presumably correct in theory. But it often happens the first of the water of condensation does not run off rapidly until the radiator is under the full pressure of the steam, when the lower end of this pipe will be covered with the water, causing the latter to rise within it by the pressure in the radiator, and ejecting it through the air-cock or valve, something that should be avoided on account of the unnecessary annoyance, if for no other reason.

In single chamber heaters, and heaters made of pipes, having free passage top and bottom, the air valve is often put near the top, the weight of the air apparently not affecting the egress.

It is better, also, to draw the air from a single pipe or a loop near the top, or, at least, not too near the lower end of the radiator, as the draft caused by the escape of the steam through a pet-cock will sometimes raise water from the base of the radiator, so that the further from the surface of the water the air can be in a vertical radiator the better.

The greatest difficulty exists in drawing the air from a *flat coil* when the return pipe does not run below the water lines, but permits live steam to enter the coil from the lower end, forcing the air toward the middle of the coil. Some steam-fitters put an air valve on a return-bend, at a point about $\frac{1}{3}$ the length of the coil from the lower end, but the result is often a disappointment. The best way in case of box coils and flat coils, is to carry their return pipes *below the water line* and place the air valve very near the return end, but so arranged as it will not draw the water from the return pipe, which it often does if placed directly on the side of the small-diameter return pipe. For this reason it is better placed on the lowest return bend near the return pipe, and any work so piped will not be likely to prove troublesome in this respect; for the current of the live steam is always from the steam to the return valve within the radiator.

The idea of the air *always* gravitating through the steam, and finding the lowest part of a heater composed of small pipes, is erroneous, *unless* the steam is let in on top, as it usually is with flat or box coils. In pipes of large diameter, air will separate of its own gravity and settle down.

In what is called the atmospheric radiator, the steam enters on top with an air hole or air cock near the bottom to let the air out, and a drain or return pipe to carry off the condensation in the bottom and deliver it to a tank without pressure. Steam enters this radiator through a very small pipe, with a nicely graduated valve, which admits any desired quantity of steam, and which fills *downward*, and permits a part, or the whole of the heating surface of the radiator to be used, and thus graduate the heat of the room by direct radiation. It may be likened to a balloon partially filled with gas, the gas always remaining in the top.*

Air and steam mix within a heater to a certain extent and at certain pressures, this mixture being of unknown weight but always greater than that of steam of the same density.

Steam at the pressure of the atmosphere, and a temperature of 212 Fahr., has a weight about one-half that of air at the pressure of the atmosphere, and a temperature of 34°; but when the air is increased in temperature about 160°, or what it would in a low

---

* These heaters cannot be used in a gravity return apparatus, but an apparatus of this kind will be described elsewhere in the book.

AIR VALVES ON RADIATORS.

pressure radiator in contact with steam, the steam will then be about two-thirds the weight of the air.

Air valves are various in design, but may be separated into four kinds; the old-fashioned pet-

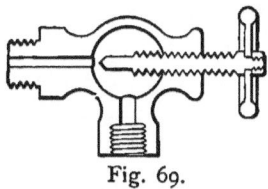

Fig. 69.

cock, the compression thumb-screw valve (Fig. 69) and the automatic air valve, working by the deferential expansion of two metals or other materials as shown in Figs. 70, 70a, 70b and 71, and the air valve similar to the last, in which there is a float that will close the valve without the intervention of expansion should water appear. Figs. 72, 72a, 73.

The pet-cock needs no explanation, and may be used on rough or factory work, but should not be used on fine or house work, for a plug cock will not stay tight on steam work, and will leak on the floors and wet the ceilings.

The compression wood handle air valve is much used, and is simply a small angle valve, with or without a stuffing-box, as shown in Fig. 69.

The automatic air valve embraces nearly as many designs as there are manufacturers of heating apparatus; but the principle used is the same in each instance, viz., the difference of expansion of any two metals that will stand the action of steam, one fo which has a greater co-efficiency of expansion than the other. The valve really becomes a metallic thermostat, which operates a little valve.

Fig. 70 shows a simple form of this arrangement; *A* being a strip of cast-iron; *B* and *b* strips of brass, set against shoulders on the cast-iron; and *C* the valve and stem passing through holes in the bar *b*, and the cast-iron *A*, and screwing into the other brass (*B*).

When heated above the temperature at which they

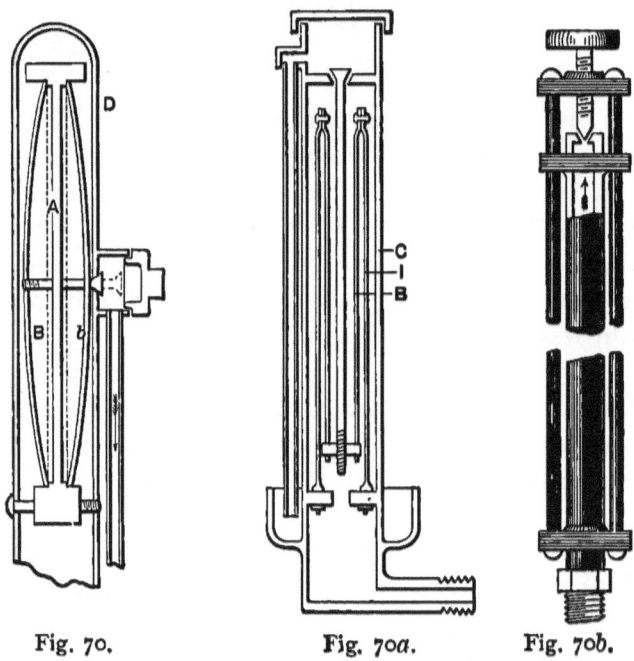

Fig. 70.   Fig. 70*a*.   Fig. 70*b*.

are fitted, the brass expands more than the iron and forms a bow shape, as shown, and draws the valve to its seat; the dotted lines show its normal position. The stem, where it screws through the brass *B*, forms a regulator, which can be adjusted with a screw driver, applied to a slot in the valve. The outside *D* may be a piece of pipe, or a casting, with a boss on the side of it, to tap a small pipe into, so as to carry the vapor away, if required.

Fig. 70a is another modification of the same principle.

The outside case C is brass, the rod I iron, or some metal with considerably less expansion than brass; the rod B of brass once more, and the center or valve rod of the same metal as the bar I. By this means two brass rods move in one direction and two iron rods relatively in the contrary direction. This is a compound differential apparatus to secure the required movement, without too great a length of tube.

Fig. 70b shows a form in which one metal only

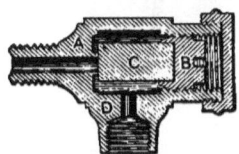

Fig 71.

(brass) may be used, the rod not being expanded as much as the case, for the reason that they are outside, and not in direct contact with the steam, though of course metals with opposite qualities are best. When the case expands, it presses on the thumb-screw at the top, forming a valve, the thumb-screw forming an adjustment. The automatic valves so far shown are old styles not much met with now, and obsolete. One of them is not a desirable feature.

The valve shown in Fig. 71 is a modern automatic air valve, made by Jenkins Bros., of New York. It was the first valve in which a plastic composition, that would stand steam and give a large coefficiency of expansion, was used. It takes up a small space, being no larger than an ordinary compression air-cock, and is suitable for a high or low pressure steam. A is the end screwed into the radiator, and B a regulating screw holding an expansible plug C, D being the outlet tapped to connect with an air pipe of the building. The plug C is of india-rubber, and other substances that have a increment of expansion, its elastic end forming a close connection with the seat A.

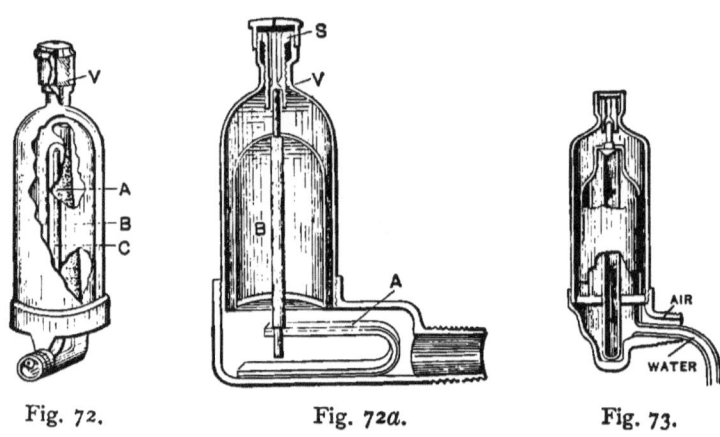

Fig. 72.   Fig. 72a.   Fig. 73.

Fig. 72 shows the Van Auken valve. It is a float valve to guard against water and has the expansion principle to close with heat. The rod A is of rubber and other substances, the expansion of which closes the valve V. Should the radiator be partially closed or the circulation of steam bad and water fill the radiator, it will pass into the chamber and lift the inverted open float B, thus closing the valve against the escape of water. A tube C is used to secure a communication between the top of the air valve and the pipe of the radiator, so water cannot be held in the case by the formation of a vacuum.

Fig. 72a (the Onderdonk) is in many respects similar to the foregoing. The movement of the valve due to heat however, is secured by the change of shape in the loop-shaped spring A. It is composed of two metals, the one with the greater difference of expansion being on the inside. The regulation of the valve V is secured by the movement of the screw seat S.

Fig. 73 is another modification of the float valve. The differential spring pinches on the cone and forces the valve to its seat. The crooked syphon pipe also assists the water to run from the case into the radiator.

CHAPTER XIV.

STEAM PIPE, SIZE, AREA, EXPANSION, ETC.

THERE are two kinds of wrought-iron steam and gas pipe—namely, lap-welded and butt-welded.

There is no lap-welded pipe smaller than 1¼ inch, though butt-welded pipe is made of all sizes, excepting those of very large diameter.

Lap-welded pipe is considered the best, although for sizes smaller than two inches there is little difference. The butt-welded pipe is the most uniform in size, though it is apt to open in the seam by twisting.

All the pipe and all fittings made in the United States and Canada are supposed to be of *standard* dimensions, so that the whole will be interchangeable.*

Occasionally in old buildings pipe is found which is known as "old gauge," which is somewhat larger than the pipe now in use.

The size of pipe is *standard*, but the standard is *arbitrary;* the inside diameter being nearest the nominal size of the pipe, which it always somewhat exceeds. Small sizes are more disproportioned (as can be seen by reference to the table of "Standard Dimensions of Wrought-iron Pipe," or to the diagram of sizes of pipe).

The threads on the ends of pipes should taper about $\frac{1}{16}$ of an inch for an inch in length of thread.

* This is not absolutely so, but they are near enough for ordinary work, and with adjustable dies the fitter finds little trouble in correcting small errors of gauge. A committee of the Society of Mechanical Engineers of which I was a member, have secured the adoption of the Biggs Standard for pipe and fitting threads since the above was first written, and it is hoped that hereafter the threads furnished by the trade will be absolutely interchangeable, the Pratt & Whitney Co. of Hartford, having commenced the preparation of the standard gauges.

96.—TABLE NO. 3.
TABLE OF STANDARD DIMENSIONS OF WROUGHT IRON WELDED PIPE FOR STEAM, GAS, OR WATER.

Inside Diameter.	Actual Outside Diameter.	Thickness.	Actual Inside Diameter.	Internal Circumference.	External Circumference.	Length of Pipe per square foot of inside surface.	Length of Pipe per square foot of outside surface.	Internal Area.	External Area.	Length of Pipe containing one cubic foot.	Weight per foot of length.	Number of threads per inch of screw.
Inches.	Inches.	Inches.	Inches.	Inches.	Inches.	Feet.	Feet.	Inches.	Inches.	Feet.	lbs.	
1/4	0.405	0.068	0.270	0.848	1.272	14.15	9.44	0.0572	0.129	2500.0	0.243	27
3/8	0.54	0.088	0.364	1.144	1.696	10.50	7.075	0.1041	0.229	1385.0	0.422	18
1/2	0.675	0.091	0.494	1.552	2.121	7.67	5.657	0.1916	0.358	751.5	0.561	18
3/4	0.84	0.109	0.623	1.957	2.652	6.13	4.502	0.3048	0.554	472.4	0.845	14
1	1.05	0.113	0.824	2.589	3.299	4.635	3.637	0.5333	0.866	270.0	1.126	14
1 1/4	1.315	0.134	1.048	3.292	4.134	3.679	2.903	0.8627	1.357	166.9	1.670	11 1/2
1 1/2	1.66	0.140	1.380	4.335	5.215	2.768	2.301	1.496	2.164	96.25	2.258	11 1/2
1 3/4	1.9	0.145	1.611	5.061	5.969	2.371	2.01	2.038	2.835	70.65	2.694	11 1/2
2	2.375	0.154	2.067	6.494	7.461	1.848	1.611	3.355	4.430	42.36	3.667	11 1/2
2 1/2	2.875	0.204	2.468	7.754	9.032	1.547	1.328	4.783	6.491	30.11	5.773	8
3	3.5	0.217	3.067	9.636	10.996	1.245	1.091	7.388	9.621	19.49	7.547	8
3 1/2	4.0	0.226	3.548	11.146	12.566	1.077	0.955	9.887	12.566	14.56	9.055	8
4	4.5	0.237	4.026	12.648	14.137	0.949	0.849	12.730	15.904	11.31	10.728	8
4 1/2	5.0	0.247	4.508	14.153	15.708	0.848	0.765	15.939	19.635	9.03	12.492	8
5	5.563	0.259	5.045	15.849	17.475	0.757	0.629	19.990	24.299	7.20	14.564	8
6	6.625	0.280	6.065	19.054	20.813	0.63	0.577	28.889	34.471	4.98	18.767	8
7	7.625	0.301	7.023	22.063	23.954	0.544	0.505	38.737	45.663	3.72	23.410	8
8	8.625	0.322	7.982	25.076	27.096	0.478	0.444	50.039	58.426	2.88	28.348	8
9	9.688	0.344	9.001	28.277	30.433	0.425	0.394	63.633	73.715	2.26	34.077	8
10	10.75	0.366	10.019	31.475	33.772	0.381	0.355	78.838	90.762	1.80	40.641	8

150 STEAM HEATING FOR BUILDINGS.

Fig. 74.—Diagram of Cross-Section of Wrought-Iron Pipe.

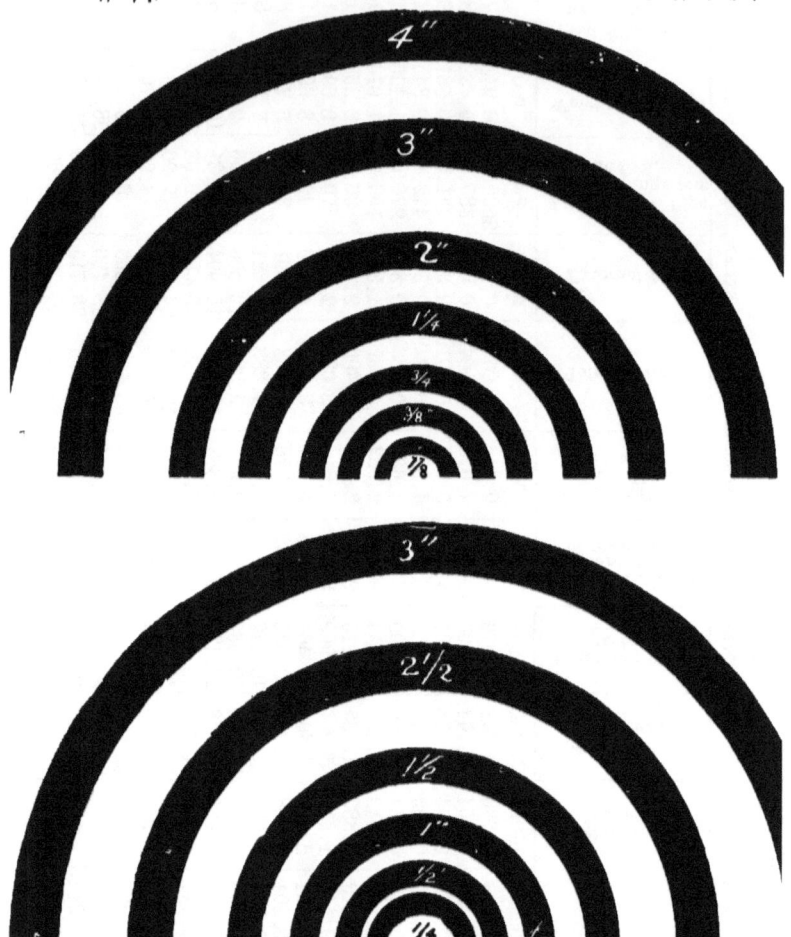

ACTUAL SIZE.

RELATIVE AREAS OF PIPE.

The young steam-fitter has not always a just conception of how the size of one pipe compares with that of another, not knowing how rapidly the area of a pipe increases with an increase of diameter.

When the diameter of a pipe is *doubled*, the area has increased fourfold, and if one pipe has one-fourth the diameter of another, it has but one-sixteenth of its area. Thus the area of the cross sections of circular pipes are to each other as the squares of their diameters.

As circles and squares always bear the same relative proportions to each other, and as either can be likened to the cross section of a pipe, the *beginner* can always find the number of times the area one pipe will divide with another, by making another, by making a square, a', Fig. 75, and calling the side of it the diameter of the smallest pipe; then around the smaller square construct a larger one, the side of it being the diameter of the larger pipe, with the corner b forming a common corner for both squares. Thus if the square a' represents a 1-inch pipe, and you draw around it a square 3½ inches on the side, and lay the larger square off into squares of the size of the smaller one, as shown, the number of the whole squares and the sum of the parts of the square within the larger square is the number of times a 1-inch pipe will go into a 3½-inch pipe.

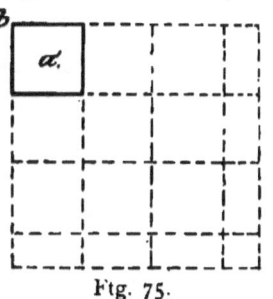

Fig. 75.

It will be seen there are nine whole squares, six half squares, and one quarter square, which equals 12¼ squares; the number of times a 1-inch pipe will go into a 3½-inch pipe.

152   STEAM HEATING FOR BUILDINGS.

To prove the above according to the rule—"*Pipes are to each other as the squares of their diameters,*" square the smaller pipe for a divisor, and the larger pipe for a dividend, and the quotient will be the number of times.

Example:

$1 \times 1 = 1.$

```
        3.5 ×
        3.5
        ───
        175
        105
     ─────────
     1.)12.25(12.25—Ans.
```

Ex.—To find how many times a ¾-inch pipe will go into a 2-inch pipe.

```
   .75 ×              2. ×
   .75                2.
   ───             ─────────
   375             .5625) 4.0000 (7.11—Ans.
   525                   │3.9375│
   ─────                  6250
   .5625                  5625
                          ────
                          6250
                          5625
                          ────
                          625 +
```

The following table has been calculated for the use of the steam and gas-fitter, and shows how many times the area of one pipe will go into another.

In practice, however, with pipes of constant lengths, more branches may be taken from a pipe than are here shown.

A large pipe, having less frictional surface for its area than a smaller, will do more work—pass more water or steam—other things, such as length, pressure, etc., being the same for both pipes.

TABLE NO. 4.

TABLE SHOWING THE RELATIVE AREAS OF STANDARD, WROUGHT-IRON GAS, WATER, AND STEAM PIPE—FROM 1/8 TO 9 INCHES, INCLUSIVE.

	1/8	1/4	3/8	1/2	3/4	1	1 1/4	1 1/2	2	2 1/2	3	3 1/2	4	5	6	7	8	9
1/8	1	4	9	16	36	64	81	144	256	400	576	784	1024	1600	2304	3136	4096	5184
1/4		1	2 1/4	4	9	16	25	36	64	100	144	196	256	400	576	784	1024	1296
3/8			1	1 7/9	4	7 1/9	11 1/9	16	28 4/9	44 4/9	64	87 1/9	113 7/9	155 5/9	256	348 4/9	455 1/9	574 8/9
1/2				1	2 1/4	4	6 1/4	9	16	25	36	49	64	100	144	196	256	324
3/4					1	1 7/9	2 7/9	4	7 1/9	11 1/9	16	21 7/9	28 4/9	44 4/9	64	87 1/9	113 7/9	144
1						1	1 9/16	2 1/4	4	6 1/4	9	12 1/4	16	25	36	49	64	81
1 1/4							1	1 11/25	2 14/25	4	5 19/25	7 21/25	10 6/25	16	21 11/25	31 9/25	40 24/25	51 11/25
1 1/2								1	1 7/9	2 7/9	4	5 4/9	7 1/9	11 1/9	16	21 7/9	28 4/9	36
2									1	1 9/16	2 1/4	3 1/16	4	6 1/4	9	12 1/4	16	20 1/4
2 1/2										1	1 11/25	1 24/25	2 14/25	4	5 19/25	7 21/25	10 6/25	12 24/25
3											1	1 13/6	1 7/9	2 7/9	4	5 4/9	7 1/9	9
3 1/2												1	1 15/49	2 2/49	2 45/49	4	5 11/49	5 30/49
4													1	1 9/16	2 1/4	3 1/16	4	5 1/16
5														1	1 11/25	1 24/25	2 14/25	3 6/25
6															1	1 13/36	1 7/9	2 1/4
7																1	1 15/49	1 32/49
8																	1	1 17/64

When *lengths* and *pressures* are equal, the discharge or quantities of steam, water or air that pipes will pass will be the ratio of the square root of the fifth power of their diameters. For instance, in the following table the upper lines represents diameters of pipe, and the second line quantities of steam:

Diameter	1"	2"	3"	4"	5'	6"	7'	8"
Quantity	1	5.65	15.6	32.0	55.9	88.2	129.6	181.0
Diameter	9"	10"	11"	12"				
Quantity	243.	316.	401.	498.				

Conversely to the above, the *diameters* of pipes, for equal *lengths* and *pressures*, will be directly as the fifth root of the square of their diameters.

To use the table of the relative areas of pipe.—Find the size of the smaller pipe, in the left-hand column, and follow it to the right, until it is under the size of the larger pipe, or *vice versa;* the number thus found is the *times* the small pipe will go into the large one.

The accompanying diagram, Fig. 76, also illustrates graphically and almost at a glance, the relative proportions of pipes, from one inch to twelve inches in diameter; the column of figures being the diameters of the pipes in inches.

The 1-inch pipe is represented by one triangle; the triangle immediately opposite the figure.

The 2-inch pipe is represented by four triangles; the three immediately opposite the figure and the one above it.

The 3-inch pipe is represented by nine triangles; the five immediately opposite, and all above it; and so on to the end.

The sum of the triangles immediately opposite the size of a pipe, and all the triangles above it, gives the square of the diameter in inches.

The number of triangles immediately opposite the size of a pipe, gives the increase in *units* of size (the unit being the area of a 1-inch pipe) over the pipe next smaller than it; and the number of triangles,

A DIAGRAM OF RELATIVE AREAS OF PIPES, FROM I TO 12 INCHES, SHOWING THE INCREASED AREA FOR EACH INCH OF INCREASE OF DIAMETER.

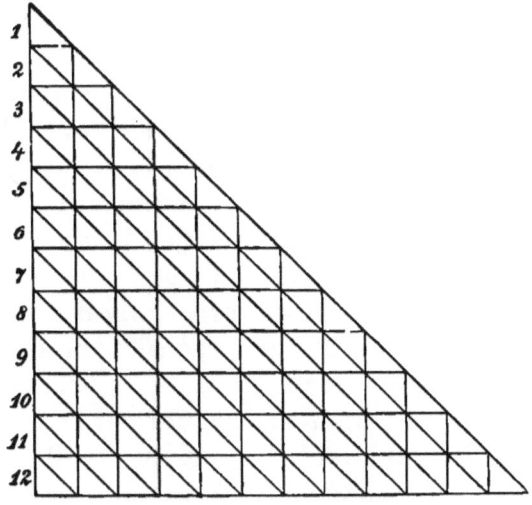

Fig. 76.

opposite the size of a pipe, with all above it, as far as the size of any other pipe, gives the increase on units for the difference between the two sizes.

It will also be seen that the increase of the area of pipes, for each inch of increase of diameter, is an arithmetical progression, whose common difference is two, the first term being one.

These simple methods may be tiresome to the advanced student, but the writer knows from contact with the workingman for 25 years that a subject cannot be made too simple.

EXPANSION OF PIPES.

In running pipe for any purpose, special attention must be given to its expansion or contraction, for nearly all leaks which occur after work is completed and tight, if not due to defective material, are caused by expansion or contraction, which has not been provided for.

When a main pipe is run close to a wall, and branches taken through holes in the wall, the holes being just sufficient for the branches to pass, the latter break off, when heated. If main pipes run some distance from the wall, the branches being unconfined near the main, even though confined near their farther ends, the *spring* of the pipe, especially if it is of small diameter, will admit of the expansion or contraction of the main in the direction of its own length without rupture. But the branches should not be confined in the direction of their length, or they will shove the main out of line, and should another branch start, directly opposite to a branch so confined, the Tee will either be pushed out of position or cross broken.

Main pipes, to look well, must be straight and should be hung so they will expand in the direction of their length, avoiding all the side motion possible and throwing the expansion of the branches in the direction of their own lengths.

Long mains should never be run very close to a wall up which risers go; for the risers admit of very little lateral movement, and all the linear expansion of the main will be thrown on the riser-connection in the form of torsion.

When a main is turned with its branch Tees looking up, a nipple and elbow can be screwed into the Tee, so as to get any desired angle in running to the wall or elsewhere. This nipple and elbow, with the pipe from the elbow, will admit of more *torsion* than a straight pipe, and in extreme cases the threads of the nipple will turn a little and prevent anything from breaking.

Special attention should be given to pipes laid between floors, or when they have to cut into floor joists or beams. They must not be confined at their ends, and their branches for 3 or 4 feet from where they leave a Tee, and should have room enough to allow for the greatest possible difference of length or change of position in the rising line.

It is common for steam-fitters to run their return pipes around cellars and basements before the concreting is done, and to allow them to be buried and cemented into this mass, which becomes as one stone, and for a time (when they do not give out upon the first turning on of steam,) must actually overcome the elasticity of the iron. But the pipe more frequently breaks or leaks, either by shoving through the threads of the fittings, or else pulling them apart, or the branches break off by having a large pipe, which may not be confined at one end, forced past them.

There is another reason why pipes shuld not be buried in floors,—namely, *lime with moisture destroys them rapidly.* Work so hid from observation is the first to give out. If connections around boilers, pumps and the like, were kept above the floor, they would probably outlast the boiler.

When hot water or steam has to be carried under ground, it must be conveyed in wrought-iron pipe, with screwed joints, or cast-iron pipe, with flanged joints. Hub and spigot pipes with leaded joints are not suitable, for it is impossible to keep them tight when subjected to much difference of temperature, as the lead expands in a different ratio from the iron, and takes a permanent set with comparatively little pressure.

Cast-iron gas or water pipes, put down in the streets, with leaded joints, it is said, will compensate in the joints by slipping; the difference on a twelve-foot length being about the $\frac{1}{50}$ of an inch for a difference in temperature of 20 degrees.

There is another explanation of this matter, which is probably the true one. It is that the pipe, being firmly held to the ground, its whole length is actually compressed equally at all parts of its length, the same as it would if subjected to compression by weight. Cast-iron is sufficiently elastic for this.

The steam-fitter should avoid using expansion joints (slip joints) when it is possible to compensate in any other way. In private houses and city buildings it can always be avoided by taking advantage of right angle turns; but frequently in long runs of pipe, in narrow passages and with pipe of large diameter, they must be used, as spring bends cannot be used unless they have considerable length, and a four or five-foot turn, on a 6-inch pipe, if the expansion was only one inch or that due to about 50 feet of pipe, would be very liable to make mischief. An eight-foot turn, on a 2-inch pipe 100 feet long, will compensate for any difference of temperature that may take place, with ordinary ranges of pressure; but on a 4-inch pipe it would in all probability break, assuming that the long run of pipe is prevented from springing sidewise.

Sometimes in running pipe through long, straight passages, if the passages have a width of about 6 feet, by frequently crossing from side to side we obtain a beneficial result; especially if it is a return pipe. The objection to this method for a steam pipe is the great

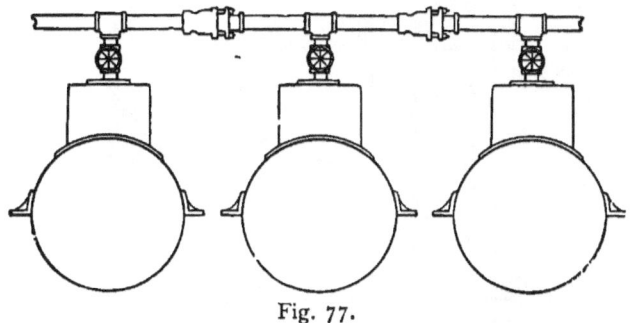

Fig. 77.

number of turns which would be required for a pipe larger than 2 inches; but when passages make one or two *right angle turns*, nothing can be better if the pipe is hung and has not to pull or push its own weight over rough surfaces, the length of pipe each way from the elbow not being sensible of any considerable bending.

When several boilers are connected together between their domes or ends, the connections should not be run "short across" from dome to dome. The pipes should be run back or forward from the domes 3 to 6 feet, and then connected across.

The reason of this is plain, when we consider that the settling of the brickwork or the expansion of the pipes will suffice to throw the weight of the boilers on *rigid connections*. For the same reasons pipes passing through the brickwork of boilers should not rest in the walls, but have large holes, covered with loose flanges, around the pipes.

Figs. 77 and 78 show plans of boiler connections, the first when using expansion joints, and the latter

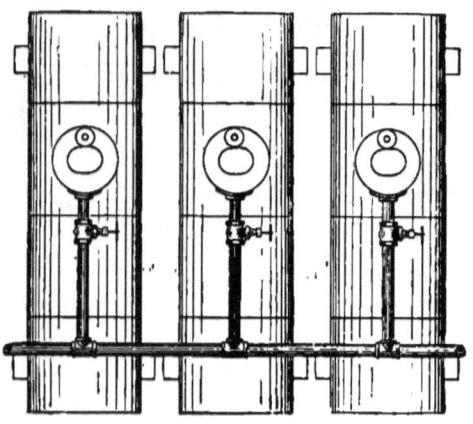

Fig. 78.

when the expansion is provided by spring, the latter being the most permanent way when properly done.

By reference to the figures it will also be seen that a slip joint only provides for a linear contraction or expansion, or a twisting motion, and does not compensate for a difference in level.

Fig. 79 shows distant rigid objects connected by a pipe, in which the expansion is provided for by the use of spring bends.

The expanding power of a 2-inch pipe, when heated

to the temperature of 100 pounds of steam, exerts a force sufficient to move 25 tons.

Cast-iron expands one one-hundred and sixty-two thousandths ($\frac{1}{162000}$) of its length for each degree Fahrenheit it is subjected to within ordinary limits, while in the solid state. Its expansion is less than wrought-iron.

Wrought-iron pipe expands one one-hundred and fifty thousandths ($\frac{1}{150000}$) of its length, for each degree Fahr. it is subjected to in any limits it can be used by the steam-fitter; and the length of the pipe in inches, multiplied by the number of degrees it

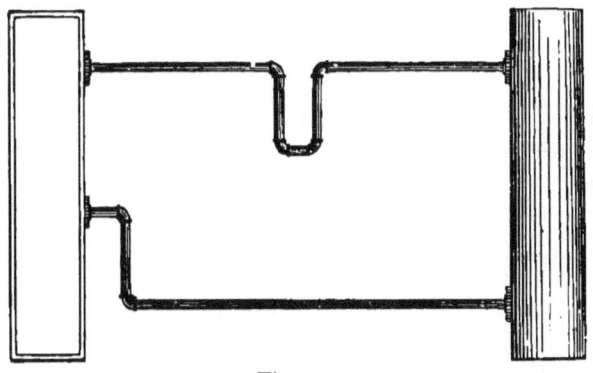

Fig. 79.

is heated and divided by 150,000, will give the expansion for that difference in temperature *in inches*, or fractions of an inch.

Example.—Find what the length of a one hundred feet of pipe will be, when heated to the temperature of 100 pounds of steam, its initial temperature being zero.

Thus, $100 \overset{\text{ft.}}{\times} 12 = 1200 \overset{\text{in.}}{\times} 328° \overset{\text{temp.}}{=} 405600 \div 150000 = 2.70$ inches. (See table.)

A TABLE OF LINEAR EXPANSION, OF WROUGHT AND CAST IRON PIPES (TO WITHIN THE $\frac{1}{100}$ OF AN INCH), FOR EACH 100 FEET IN LENGTH, AT TEMPERATURES AND PRESSURES MOST FREQUENTLY REQUIRED BY THE STEAM-FITTER.*

WROUGHT IRON.

Temperature of the Air, when the pipe is fitted.	Length of pipe when fitted.	Length of pipe when heated to			
		215° or 1 lb. of steam.	265° or 25 lbs. of steam.	297° or 50 lbs. of steam.	338° or 100 lbs. of steam.
Degrees, Fahr.	Feet.	Feet. Ins.	Feet. Ins.	Feet. Ins.	Feet. Ins.
0	100	100 1.72	100 2.12	100 2.31	100 2.70
32	100	100 1.47	100 1.78	100 2.12	100 2.45
64	100	100 1.21	100 1.61	100 1.86	100 2.19

CAST IRON.

0	100	100 1.59	100 1.96	100 2.20	100 2.50
32	100	100 1.36	100 1.65	100 1.96	100 2.27
64	100	100 1.12	100 1.43	100 1.73	100 2.00

*Calculated for Regnault's temperatures and Lavoisier and Laplace's difference of expansion.

Rolled wrought-iron expands the 150000th part of its length to each degree Fahrenheit it is warmed, and soft irons and forgings expand slightly more, whereas steel as a general thing expands somewhat less. For all practical purposes, such as the architect or the steam-fitter have to deal with in heating, steel and iron may be assumed to expand substantially the same, varying slightly with the conditions of its hardness, etc.

CHAPTER XV.

SIZE OF MAIN PIPES FOR LOW PRESSURE STEAM HEATING APPARATUS, AND WHY SUCH SIZES ARE NECESSARY.

No gravity heating apparatus is perfect unless it heats thoroughly at all pressures; unless the water of condensation runs back and into the boiler at all pressures; unless it is noiseless under all ordinary conditions, so that the duty of the person in charge is simply to take care of the fires and see there is always sufficient water in the boilers.

The fitter in all probability knows that a gravity apparatus requires larger pipes than any other system, and thus he can take it for granted the size of piping sufficient for a gravity apparatus will be enough for any other description of work.

As this pipe is principally devoted to the heating of buildings and blocks, which have their own boilers, situated either in the buildings or near to them, the rule mentioned hereafter is intended for determining the size of main pipes for gravity apparatus for all ranges of pressure, or where pressure is required throughout an apparatus that is nearly an *initial* pressure—that is, a pressure the same as within the boiler to within, say, ½ pound, or at most 1 pound, at the remotest part of the apparatus,—as with an automatic direct return steam-trap job or an exhaust steam job.

With high pressure steam, which is allowed to expand through a building and eventually escape through atmospheric traps or tanks, a very much smaller piping will do; but the waste of heat is sometimes very great with traps which discharge into an open tank or to atmosphere. The difference in favor of a gravity apparatus, or an apparatus working properly, with direct return traps or pump-governing system, can always be estimated at 15 per cent. over apparatus which permits the water to escape and thus loses it, with the heat it contains; and when traps are neglected (which is the rule) it may reach 30 per cent. of all the heat.

This is not an assertion in the interest of direct return, or one which cannot be verified, as the following will show.

When water is returned to the boiler at a temperature of 180° (the ordinary temperature of water from gravity apparatus), it requires 1,000 heat units to make one pound of it a pound of steam, and in condensation to water again, and returning it to the boiler at 180°, it loses just 1,000 heat units; which have all been utilized within the building. Thus every unit of heat, added to the water, has been realized, and it represents *the maximum economy possible in steam heating;* the power required to put the water back being at a minimum—*i. e.*, gravity. In the case of an apparatus that wastes its return water, and has to pump water from the waterworks at a temperature of 40°, it has to add to every pound of water converted to steam, 1,140 *units*, and gets only 1,000 from it, when the water is cooled to 180° (a very low temperature, by the way, for ordinary traps to expel water at). Thus, for every 1,140 units added to the water, 140 are lost, or over $12\frac{1}{4}$ per cent. When the pressure in the radiator is 40 pounds and the water passes the trap at a temperature corresponding to that pressure (285° Fahr.), more heat is allowed to waste, as there are 1,140 units required to raise fresh water at a temperature of 40° to steam at 40 pounds, and only 902, utilized in cooling to 285°, the temperature of water at 40 pounds, which leaves 245 units unaccounted for, or a *loss* of more than $21\frac{1}{4}$ per cent.; and this does not take into consideration the heat lost in pumping water into the boiler.

The power necessary to pump water into a boiler is a little over $\frac{1}{2}$ of 1 per cent. of all the power obtained from the steam, and for common pumps it should be placed at not less than 1 per cent. of all the steam. In fewer and plainer words, 1 per cent. of the steam furnished by the boiler is a fair allowance to be chargeable to putting the water into the boiler again, when it has to be done by a steam pump.

If the water from traps, discharging at 40 pounds pressure, is saved in one tank, and pumped into the boiler again, then the condensed water, after being received into the tank, will have a temperature of about 200°. But it will be said the water escaping from a trap at 40 pounds pressure had a temperature of 285°, hence the water will be received at that temperature if the tank is kept under pressure; and this is true of the modern pump and governor method of returning condensed steam. But where it is necessary to have a tank open to the atmosphere (with either an overflow pipe or a vapor pipe) to receive the water, and as water at a pressure of the atmosphere cannot have a temperature above 212°, the difference will escape in vapor, or low pressure steam, through the vapor pipe; and if we have a *tight tank* without traps, we must have *as large* pipes, or nearly as large pipes, to get water to gravitate to the tank as are required for boiler gravitation; so that when the difference in a cellar or basement will permit, it is better to put the water of condensation directly into the boiler than to trap or pump it. But to return. The temperature of the water in the open tank we will take at 200, and to raise a pound of it to steam will require 979 units, and only 894 units of it will be realized in cooling, if it passes the trap at a temperature corresponding by getting into a condition fit to remain in the tank—this is over $8\frac{1}{2}$ per cent., to which add 1 per cent. for pumping the water back, and the sum will equal $9\frac{1}{2}$ per cent; but should the water be lost each time and fresh cold water be supplied, it will equal $21\frac{3}{4}$ per cent.

Thus it will be seen, it is poor economy to use small pipes and resort to tanks, traps, pumps and other contrivances, to get water back, when the *price of a steam pump* expended on larger pipe is frequently sufficient to get the water back, and obtain an effect, which so far as the heating surface is concerned, will give the *maximum duty*, and do away with one source of continual expense, as well as the loss of heat occasioned by such irregular means. Twenty-five or thirty years ago it was excusable, in some, because it was not then generally known that water could be returned *at all pressures;* but now it is unpardonable, when the circumstances of the case, position of building, etc., will admit of doing better. Furthermore, it should be the duty of the architect to provide, if possible, for *direct return* in the general planning of buildings, at least for direct heating, when there is little or no exhaust steam to be taken care of, such as in public institutions and private residences, particularly the latter. Since the use of exhaust steam has become general in the big office buildings of the country it has become necessary to pump and return condensed water by mechanical means. It was the custom some years ago to take steam from the boilers to run the elevator pumps and other engines of the building, and waste the exhaust steam and take live steam for the heating apparatus from the same boiler, though by another pipe. This, however, is now all changed in this class of buildings. New York City buildings, and big buildings throughout the country to-day, use much more steam for their elevator service and electric light service than can be condensed in the heating apparatus, and no designer or engineer with any regard for his reputation will now waste the exhaust steam, but will turn it into the heating apparatus and condense it as far as it will go, often having to waste a considerable quantity of it through the exhaust pipe, not being able to condense it all; though sometimes having to add a little live steam to make up a deficiency when the exhaust is not sufficient. Apparatus of this kind, of course, cannot be on the gravity system, though usually the size of the pipes given for gravity work seem to suit this class of work better than any other, for the reason that the pressures of steam are so small that it will require pipes equally as large as for gravity work, though perhaps not always run in the same manner.

The Exhaust System of Steam Heating will be treated of more fully in a later chapter.

The conditions first cited, however, are often found in factories and country workshops where the heating pipe is taken from the boiler that makes steam for the engine and carried to heat the factory or office. The condensed water and discharge from the traps is often found running into the sewers or drains of the building, or into a creek if one is near by; but the waste from the loss of the water, etc., does not stop here. It will be found occasionally there are no traps on the pipe, the whole thing being controlled by a valve, though more frequently it will be found that the traps are inoperative and allowed to waste the best way they can, in which case there is often more steam going to waste than is required for warming that part of the building. We are, however, going away from our subject, and the reference above given was here introduced as a most conclusive argument for properly arranging the size of the pipes and the system of heating to be used at the outset.

There is no very definite rule among those who do steam-heating, or there certainly was not before the first editions of this book were in print, by which they may determine the correct size of pipes; hence much confusion and many failures, to the general injury of the trade, though of recent years a great improvement has been made in this respect. Those who make a specialty of heating soon find they must use large pipes, and they generally adopt some arbitrary unit, such as to allow the size of a ¾-inch pipe to each radiator; a half a square inch in the cross section of the horizontal main to each 100 square feet of heating surface or to each radiator; and the area of a *one-inch pipe* to each 100 square feet of heating surface. The latter I early adopted as a rough rule for my workmen, the only one that in my experience was ample.

This rule also compares very nearly with deductions made from the steam pipes of certain buildings throughout the country which are considered repretative pieces of work, and have proved themselves ample when the greatest cold prevailed.

Thus, the area of a *one-inch steam pipe*, 0.7854 of a square inch, may be taken as the *unit;* and it serves very well, as by simply squaring the diameter of a pipe *in inches*, you have the number of 1-inch pipes, or *units*, or hundreds of square feet, of pipe or plate surface, the main pipe will surely supply steam if the pipes are not of too great a length. Thus a 3-inch pipe will supply steam for 900 square feet of heating surface, when subjected to the greatest condensation possible within buildings, and still not raise the water line in the pipes to any appreciable extent.

It happens that the area of a one-inch pipe (0.7854 of a square inch) makes a very satisfactory unit. The diameter of a steam pipe always increases directly as the square root of the heating surface, and according to the arbitrary unit here adopted, *the diameter of the pipe in inches*, will be exactly *one-tenth* of the square root of the heating surface, in feet. Thus, when you find your heating surface, extract its square root, in feet, and call one-tenth of it the diameter of the main, *in inches*.

This is on the assumption that the mains increase in length in a certain proportion to their diameters. For instance, assuming the 2-inch pipe to be not over 50 feet in length, the 2½ about 75 feet long, 3-inch 100 feet, a 4-inch pipe about 150 feet, a 5-inch pipe about 200 feet, and each successive size about 100 feet longer than the one that preceded it.

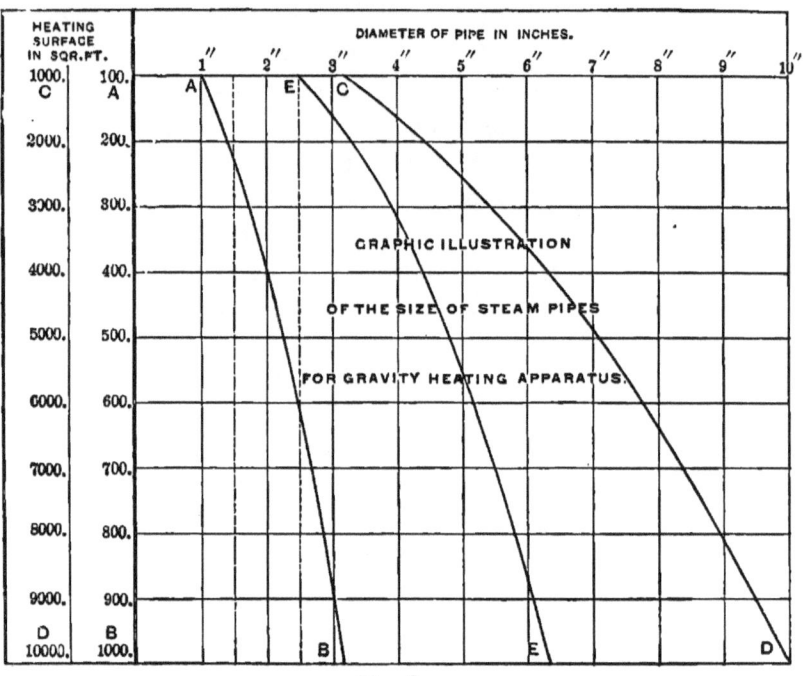

Fig. 80

These are about the conditions found in long, low buildings, such as insane asylums, hospitals, depots, etc., and the accompanying diagram, Fig. 80, may be used without much error for such buildings. It illustrates the subject at a glance, and gives the size of main pipes for surfaces, from 100 to 10,000 square feet.

The ordinates of the curve, AB, gives diameters corresponding to the square feet of heating surface in the column marked AB.

The ordinates of the curve CD bear the same relation to the column CD as the curve AB bears to the column AB, and shows the size of pipe for heating surfaces from 1,000 to 10,000 square feet.

It will be seen that 1,000 at the head of the column AB corresponds to 1,000 at the bottom of the column CD, the ordinates of both curves agree near the 3-inch pipe.

Example.—Required the size pipe, for 600 square feet of heating surface. Find 600 in the column, and follow the horizontal line to where it crosses the curve AB; then follow the nearest perpendicular line to the nearest size of standard pipe *above* the line, in case it should not come exactly on a standard size; in this case it is a little below $2\frac{1}{2}$-inch pipe, which size should be taken.

Pipe of less than $1\frac{1}{2}$ inches diameter should not be used horizontally in a main, unless for a single radiator connection.

All this, of course, is an empirical rule founded on good practice. The rule, however, is closely correct for practical conditions, and for all sizes of pipes up to about six inches in diameter is not likely to give results much greater than claimed for it. In practice in small mains, say smaller than three inches or when they appear to be slightly longer than usual, I generally take the next size larger pipe than will be given by this rule, and I look on the rule as furnishing sizes that are just ample for gravity work without being in the least unnecessarily large.

This rule may be used to determine the size of the steam pipe in radiator connections by *increasing the pipes one size*, to give them a practical magnitude, to overcome loss by short turns, etc. Main pipes should not decrease in size, according to the area of their branches, but should be proportioned at their various stages by the same rule as for determining the size of the main the first time. The same is true of the large branches. Find surface they have to supply steam for, and proportion them as you would a main close to the boiler; figuring their own surface as radiating surface unless they are to be covered. When the mains and distributing pipes are to be covered with some good nonconducting material, their surface need not be figured as against their size, but when they are excessively long, or exposed in cold places, their surface should be considered.

Of course it is not necessary to use main pipes of as great a diameter as given above, if the mains and coils are very much above the boiler; but for cellars or basements 10 feet, or under, in height, it will not be found too large or wasteful. Discretion also can be used in the use of this rule, when pipes run 4 inches or over, as 2,000 to 2,500 feet of direct heating surface may be taken from a 4-inch pipe under favorable circumstances, provided its branches follow the rule closely. A 6-inch pipe will be ample for 5,000 feet of surface under good conditions, high basements, and 10-inch for 15,000, if not too long, or rather, if short, and such as are found in city buildings that are high but on a small ground area.

For pipes of equal short lengths the increase of diameters would be in the ratios of the fifth root of the square of the radiating surface, as mentioned before, which would call for about a 2½-inch pipe for 1,000 feet of surface, a 4-inch pipe for 3,000 square feet, a 5-inch pipe for 5,500 square feet, and a 6-inch pipe for 8,700 square feet; and this will be ample for short lengths of straight pipe without elbows, but not for the ordinary ramification of pipes of a building. The line EE shows diameters for constant lengths.

Between the lines EE and CD of the diagram are to be found the true conditions for high buildings with short mains, such as are to be found in New York and other large cities.

If the line EE is used, add at least one inch to the diameter of the pipes to overcome the resistance caused by the short turns of valves and elbows. Then, as the buildings decrease in height and increase in horizontal area, go nearer to the line CD.

These sizes will do for either exhaust steam work or gravity work under fair conditions.

CHAPTER XVI.

STEAM.

TEMPERATURES of steam, according to the different formulas, all agree at the atmospheric pressure, but as the pressures become high, they vary slightly: Regnault and Rankine are nearly alike, while the experiments of the Franklin Institute are about five degrees higher for 75 pounds apparent pressure.

The technical terms, used about steam by writers, and the expressions in vogue amongst steam-fitters, want some explanation to make them clear, as many of them are synonymous, and the fitter does not always know what is meant.

Temperature.—A condition of a body which in all cases determines its readiness to part with heat to surrounding bodies or to receive it from them. It is usually expressed in English and American books in degrees of Fahrenheit's scale. The *heat of steam* as distinguished from the *heat in steam*.

Pressure—Is the force of steam, usually expressed in pounds per square inch, and called "elastic force," "expansive force," "tension," and "elasticity," etc., are synonyms.

Density.—The weight of a cubic foot of steam compared to a cubic foot of water. Syn.—The weight of water necessary to form the steam.

Maximum density of steam.—The density of steam when it is neither superheated nor laden with particles of water mechanically. Syns.—Steam at its maximum density is called dry saturated steam or dry steam.

Superheated steam.—Steam expanded by heat, or under an increased pressure, due to an increase of heat, without the addition of water. The steam is, of course, dry beyond the point of maximum density.

Wet steam.—That containing water carried by force of ebullition, and held in the steam by the rapidity of evolution, when the steam space of a boiler is not large enough.

Foaming.—A condition differing from wet or saturated steam, by having an excess of some foreign substance in the water, causing it to foam and seem lathery, and which appears to give the water in the boiler a bulk above what would be due to the pressure, by retarding the proper separation of the steam and raising the whole mass of water into a froth. Syns.—Priming; drawing water.

This condition differs from "priming." Priming is generally a mechanical effect, while foaming is the results of foreign substances in the water. Foaming and priming are often confounded. To prove whether the boiler is foaming or priming it is only necessary to shut the main valve on the boiler tightly for a moment, when, if the boiler is foaming, there will be no change of level in the water glass, whereas if the boiler is priming through a mechanical effect, the water will at once seek its proper level and remain constant until the valve is again opened. Foaming may cause priming by sending water out of the boiler through the steam pipes; in other words, priming may take place with dirty water when it will not take place with clean water. Priming and foaming may occur together.

Volume.—The space occupied by a given quantity of water, should the water be converted into steam. The *relative volume* of steam decreases as the pressure increases. Syns.—Relative volume; bulk for bulk.

Specific gravity of steam.—The *weight* of its volume compared to the same *bulk* of water, air, or any other substance it is contrasted with. Syn.—Density.

Specific heat of dry-saturated steam.—The *heat* of a given *weight* of steam compared to a given *weight* of water or any other substance it is contrasted with. The heat necessary to raise the temperature of a pound of steam one degree while it remains continually at its maximum density.

The annexed table gives the *apparent* pressure of steam from atmosphere to 100 pounds in *pounds* per square inch, *absolute* pressures in *inches* of mercury, and *temperatures* in degrees Fahrenheit (to within one-half degree), according to Regnault, the *volume* being calculated.

TABLE NO. 5.

ELASTIC FORCE, TEMPERATURE AND VOLUME OF STEAM.

ELASTIC FORCE.		Temperature of Steam corresponding to its Pressure.	RELATIVE VOLUME	Average Rise of Temperature for one lb. Pressure for each 10 lbs.
Apparent Pressure of Steam in lbs. per Square Inch.	*Absolute* Pressure in Inches of Mercury.		Vol. of Steam compared to Vol. of Water.	
0	30.0	212.0	1710.0	
1	32.03	215.5	1612.0	
2	34.07	219.0	1523.0	
3	36.11	222.0	1442.0	
4	38.15	225.0	1372.0	
5	40.18	227.5	1312.0	2.8
6	42.22	230.0	1248.0	
7	44.27	232.5	1194.0	
8	46.30	235.0	1168.0	
9	48.33	237.5	1103.0	
10	50.37	240.0	1061.0	
11	....	242.0	....	
12	....	244.0	....	
13	....	246.0	....	
14	....	248.0	....	
15	60.56	250.0	895.0	1.75
16	....	252.0	....	
17	....	253.5	....	
18	....	254.5	....	
19	....	256.0	....	
20	70.75	257.5	718.0	
21	....	259.0	....	
22	....	260.5	....	
23	....	262.0	....	
24	....	263.5	700.0	
25	80.91	265.0	684.0	1.5
26	....	266.5	....	
27	....	268.0	....	
28	....	269.5	....	
29	....	271.0	....	
30	91.12	272.5	614.0	

TABLE No. 5—*Continued.*

ELASTIC FORCE.		Temperature of Steam corresponding to its Pressure.	RELATIVE VOLUME	Average Rise of Temperature for one lb. Pressure for each 10 lbs.
Apparent Pressure of Steam in lbs. per Square Inch.	Absolute Pressure in Inches of Mercury.		Vol. of Steam compared to Vol. of Water.	
31	....	274.0	....	⎫
32	....	275.5	....	
33	....	277.0	....	
34	....	278.5	....	
35	101.81	279.5	558.	⎬ 1.3
36	....	280.0	....	
37	....	282.0	....	
38	....	283.0	....	
39	....	284.5	....	
40	111.5	285.5	510	⎭
41	....	286.5	....	⎫
42	....	288.0	....	
43	....	289.0	....	⎬ 1.15
44	....	290.0	....	
45	121.7	291.0	470.	
50	131.88	297.0	435.	⎭
55	....	302.0	....	⎱ 1.0
60	152.25	307.0	390.	
65	....	311.0	....	⎱ 0.8
70	172.43	315.0	343.	
75	....	320.0	....	⎱ 0.8
80	193.0	323.0	305.	
85	....	327.0	....	⎱ 0.7
90	213.38	331.0	283.	
95	....	334.0	....	⎱ 0.65
100	233.76	337.5	260.	

When the pressure in inches of mercury is not given, multiply the *apparent* pressure in pounds per square inch by 2.0376, and the answer will be the *inches of mercury above atmosphere;* or that which an old fashioned mercury column would show.

Example.—10 pounds × 2.0376 = 20.376 inches of mercury.

If the *absolute* pressure is required, add 30 to the above. (20.37 + 30 = 50.37. See table.)

STEAM. 179

When the *volume* of steam is not given, add 459 to the temperature of the steam; multiply the product by 76.5, and divide by the absolute pressure in inches of mercury; the answer is the *volume*, or number of cubic feet a cubic foot of water will occupy when made into steam at the pressure required.

Example.—Required the *volume* for 10 pounds pressure, temperature 240° Fahr.—$240 + 459 = 699 \times 76.5 = 53473.50 \div 50.37 = 1061.9$ (see table).

To find what a cubic foot of steam will weigh at different pressures, divide 1000 by the *volume*, corresponding to the required pressure, and the answer will be the weight in *ounces*.

Example.—What will a cubic foot of steam at maximum density weigh, at 40 pounds per square inch?— Volume $510 \div 1000 = 1.96$ oz.

To find the number of cubic feet of steam a pound of water will make at the different pressures.—Divide the weight of a cubic foot in ounces (as above) into 16, and the answer will be the volume in *cubic feet to the pound*.

Example.—How many cubic feet of steam, at twenty pounds pressure will one pound of water make?— Volume $718 \div 1000 = 1.39 \div 16.0 = 11.5$ cubic feet to the pound of water. (See Diagram of dry saturated steam.)

To find the weight of steam necessary to raise a given quantity of water a certain number of degrees. Subtract the lowest temperature of the water from that to which it is to be heated for a dividend,— subtract the highest temperature of the water, from 1147 for a divisor, and the quotient from these will

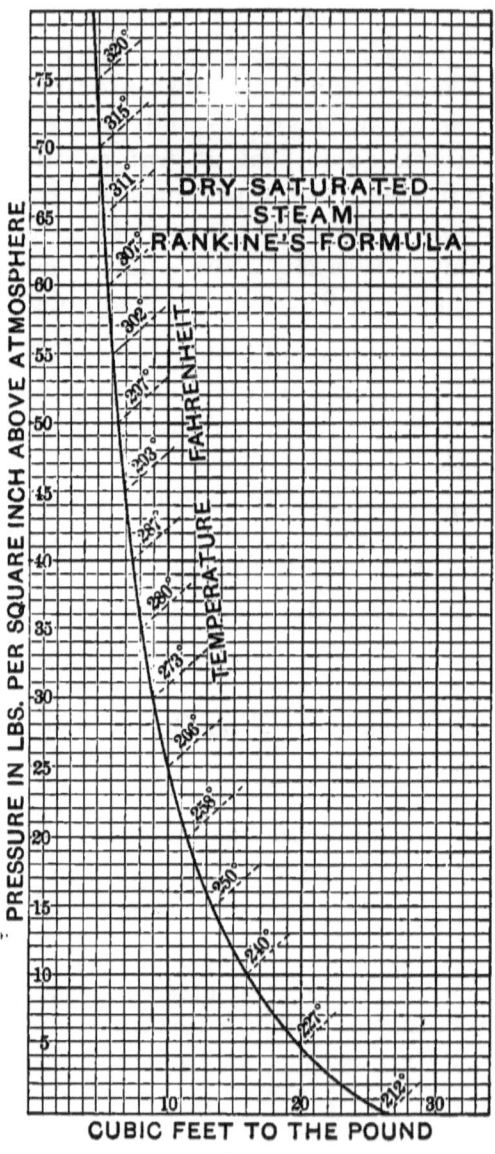

Fig. 81.

be the weight of the steam compared to the weight of water.

Example.—Find the weight of steam necessary to raise water from 75° to 190°. Thus 190° − 75° = 115, for a dividend 1147. − 190 = 957 for a divisor, 957 ÷ 115 = .12 or $\frac{12}{100}$ the weight of the water.

To find the weight of water, a given weight of steam will heat.—Proceed as above, only *transpose* the divisor and dividend.

Example.—115 ÷ 957 = 8.32 times the weight of the steam.

The accompanying diagram, Fig. 81, of Rankine's formula has been modified to commence at the atmospheric pressure — 15.7 of the absolute scale, being one pound here, and shows at a glance the cubic feet of steam to the pound weight of water at the different pressures, as well as the temperatures corresponding to the pressure.

CHAPTER XVII.

HEAT OF STEAM.

THE unit of heat is the amount required to raise the temperature of one pound of water one degree Fahrenheit, and is the standard measure of values used in all calculations pertaining to heat.

The equivalent in work of the unit of heat is the raising of approximately 772 pounds avoirdupois one foot high, and is called the *mechanical equivalent of heat*. The expenditure of one heat unit will raise the temperature of 52.5 cubic feet of dry air from 32° to 33° Fahrenheit. In very accurate calculations the moisture in the air should also be considered, but as it is a variable quantity it cannot be considered here. In approximate calculations, when estimating the heat necessary for the air of a building, it may be assumed that 50 cubic feet of air at the average outside winter temperature of air, warmed 1 degree Fahr., is the equivalent of the *heat unit*, and it can be taken as a constant.

Sensible and latent heat.—Steam has a temperature corresponding to its pressure, as given in the table, and this *apparent temperature* is known as the *sensible heat of steam*. But it is found that steam contains more heat than a thermometer will show, heat that can be made manifest in the warming of air, water, etc., and by warming a very much larger quantity than would appear by a comparison of the temperature of the steam with the ordinary temperatures of water. This *extra* heat, which is not sensible to the thermometer, is called the *latent heat of steam*.

When a solid becomes a liquid, or a liquid becomes a vapor, heat is absorbed by the body in greater quantities than is necessary to raise it to the temperature at which the change of state occurs. This *latent* heat does work in the destruction of the force of cohesion and other occult changes which take place, and must be absorbed from some substance. In the case of water and steam in a boiler, it comes from the fuel during combustion, and when a pool of water is vaporized in the street the heat comes from the sun directly, or from the earth, air, etc., indirectly. When steam or vapor is condensed, this same quantity of heat that was received, no matter where, is again given off to any substance within its influence, air, water, etc., colder than itself; and it is this property, to convey *more* heat within ordinary controllable temperatures than any other substance which makes water and its vapor so valuable.*

It takes as much heat to melt a pound of snow from a temperature of 32°, to water at 32°, as would warm a pound of water from 70° to 212°. This heat is absorbed by the water in changing from a solid to a liquid, and must be given off again before the water can be frozen.

* Water has the greatest specific heat of any known substance, with two unimportant exceptions; one of them being the principal component of water.

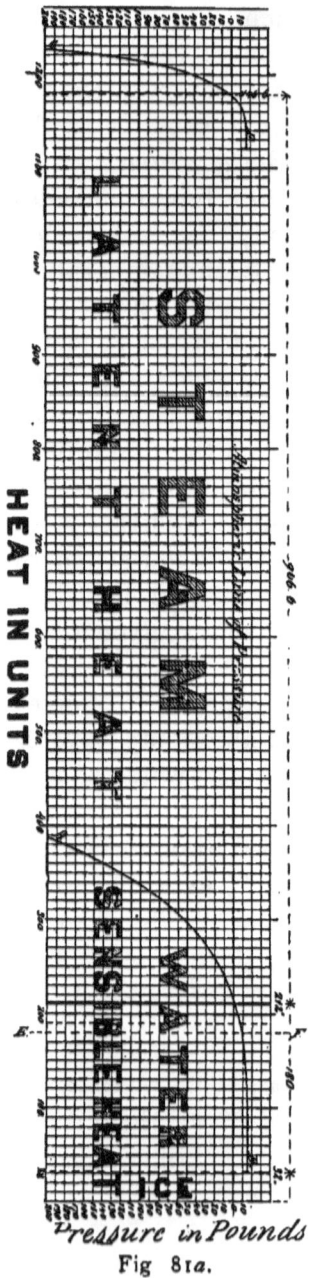

Fig 81a.

With water from the temperature of ice (about 32° Fahr.) to 212° under the pressure of the atmosphere, there is no heat made latent in confinement, each pound of water receiving only 180 heat units; but in the conversion of one pound of water at 212°, to steam at 212°, it receives 966 more units of heat; enough to warm $5\frac{1}{2}$ pounds of water from 32° to 212°, or to cool 9 pounds of iron from redness to zero. This heat is the *latent heat*, and the real *thermal* value of the steam.

The *sum* of the sensible and latent heat of steam is nearly the same for all pressures. At atmospheric pressure the sensible heat is 212°, and the latent 966°.6, giving 1,178°.6 as the total heat. At 100 pounds the sensible heat is 337°.5, and the latent 874°.8, giving 1212.3 as the total heat, the difference being 33.7, but this *difference* is not manifest in the heating of water when the steam is not allowed to expand to atmospheric pressure in cooling, as it expands itself in force, which would be manifest in an engine. Steam allowed to expand to its full volume against the pressure of the atmosphere exerts nearly the same force as if expanded against the piston of an engine. Actually, the extra heat is carried out of the boiler, at high pressures, being another form of heat *made latent*, the *extra* units remaining in the steam in the form of stored energy until it is expanded.

The assertion, therefore, that *the total heat of steam is the same for all pressures* is nearly correct in making calculations on warming, as it is presumed the steam is expanded to atmosphere in use; but when high pressure steam is condensed to water under its own pressure, then the total heat of the steam can not be considered, as the latent heat for the given pressures only, is available. The total heat of steam, however, increases, according to the experiments of Regnault, as the pressure advances, and the annexed diagram, Fig. 81a, has been constructed from the tables of Regnault, to show the way in which this increase occurs.

It also shows the number of *units* of latent and sensible heat of steam compared with each other, the ordinates of the curves AB showing the sensible heat, from one pound pressure to 200, counting from the *line marked zero*, or counting from any other imaginary line, as 32° (ice), or from the line EF, which may be taken as the temperature of return water. The difference between ordinates of the curve AB and the curve CD, gives the *latent* heat of steam for the different pressures noted. The difference between the ordinates of the curve CD and the *constant line* 1146.6, shows the increase of the *sum* of the heat above the constant 1146.6.

A pound of water converted to vapor in the open air, or a pound of water vaporized from clothing in a drying-room, requires very nearly the same heat as would be required to evaporate one pound of water to steam in a boiler, and for all practical calculations it may be taken as the same. Thus, the weight of steam necessary to dry clothing or to evaporate water in any kind of cooking apparatus, etc., can never be less than the weight of the water driven off in steam or vapor; and of necessity it will be greater, to supply the loss by radiation, or in warming the fresh air of a drying room (which must be changed as often as it becomes saturated), and for other reasons.

Equivalents of heat.—The heat necessary to warm a pound of water at *mean temperature* (39° Fahr.) one degree (*the heat unit*), will warm very nearly four (3.94) pounds of air, one degree; $2\frac{1}{10}$ pounds of vapor of water, one degree; 9 pounds of iron, one degree; and very nearly 2 pounds of ice, one degree.*

The heat necessary to convert one pound of water from the temperature of feed water, or return water, at 178°, to steam at one pound pressure (or to any pressure, not noting the slight increase for high pressures), is 1,000 *heat units*, and will heat 52,500 cubic feet of dry air one degree, above 32° Fahr., or 5,250 cubic feet of air 10 degrees; or 525 cubic feet of air 100 degrees, making no allowance for the expansion of the air, which will increase the bulk $\frac{1}{5}$ for a difference of 100 degrees; in other words, the 525 cubic feet will be increased to 630 when heated 100 degrees, and the 5,250 will be increased to 5,360 or

---

* It must not be confounded with melting the ice, but refers to changing the temperature of ice below 32 degrees.

$\frac{1}{48}$ of its bulk for a rise of temperature of 10 degrees.

The heat necessary to warm one cubic foot of water, from the temperature of the return water to steam, is capable of warming 45,572 cubic feet of dry air from zero to 72°, but if the air absorbs only 5 grains of vapor of water for each cubic foot—as from clothes in drying room, or wet walls of a building—it will be equivalent to the fall of the temperature of the air to 34.5. If the moisture is already in the air, and has only to be warmed (superheated), it will not be equal to the cooling of it more than one and a half degree.

One *grain* of water vaporized is equivalent to cooling from 7.5 to 8.6 cubic feet of air one degree, according to the initial temperature of the moisture, and is a *constant;* but 1,000 grains of vapor *already* in the air, warmed *any* number of degrees, cools only $3\frac{1}{2}$ to 4 cubic feet of the air the same number of degrees.

When water is evaporated at the expense of the heat of the air, it makes a large factor, which cannot be overlooked; but vapor already in the air, when warmed along with the air, forms a small factor and is not of much practical consequence.

The above facts explain why a new damp building often proves difficult to warm for the first few months, while later on there is often an over-abundance of heat from the same apparatus.

CHAPTER XVIII.

AIR.

Air is a *mixture* whose parts are not chemically combined, consisting of about 77 per cent. of *nitrogen* and 23 per cent. of *oxygen*, by weight, when considered *pure, i. e.*, when it is in the condition best suited to support animal life. It also contains from about $\frac{2}{10000}$ to $\frac{2}{1000}$ of its volume of carbonic acid gas, according to circumstances and location, and some watery vapor, and is capable of absorbing any other gas or vapor, to a certain extent, distributing them throughout the whole atmosphere, by what is called *the law of gaseous diffusion*, a property which gases have of mixing and diluting, which prevents gases of the most opposite specific gravities from stratifying for any considerable time. Prof. Youmans says this effect will be produced even through a membrane of india-rubber; carbonic gas rising and mixing with hydrogen, though twenty times heavier. Thus, exhaled air, and air contaminated in any other way, is perpetually made respirable by diffusion.

This property is of the utmost importance to air, for if its elements were to become separated, or an added noxious gas to remain separated from the mass, deadly gases would be the result in all unventilated places in a very short time. It frequently happens in mines and wells, where the entrance is small, and there are not sufficient disturbing influences, that injurious gases become abundant, the diffusion being too slow.

In confinement, air may have its oxygen increased or diminished; an increase of 2 or 3 per cent. causing fever, and a diminution of 3 per cent. causing death, if the carbonic acid gas from the lungs is exhaled into such air and the air inhaled afterward.

The amount of *fresh* air necessary for respiration for an adult is often stated to be about 300 cubic feet in 24 hours. This general statement, however, is misleading, and the idea that is intended to be conveyed is, that an average individual requires the oxygen of about 300 cubic feet of air in each 24 hours to support life. Air cannot be breathed in such a manner that all its oxygen will be extracted.

Air in rooms is likely to be breathed again, in a more or less degree, and as it is vitiated by moisture from the skin and lungs, and by other means well known to people of ordinary intelligence, 300 cubic feet per hour is far too little to provide in ordinary ventilating; and then not with the expectation of keeping the air fairly pure, but rather in a state which will not be injurious, even if it receives no other contamination than that from the body in health.

Hospitals should be supplied with ventilating apparatus capable of supplying at least 3,000 cubic feet of air per hour to each patient, with means to double or quadruple the quantity by forcing it (as with a fan), in times of contagious disease, or in very oppressive weather.

School and class-rooms should have at least from 1,500 cubic feet of fresh air per hour per child, for large children or the higher classes, to 1,000 cubic feet for small children, ranging between as the classes advance. This is considered a fair allowance in view of the practical difficulty of admitting so much air in the aggregate without making drafts. The Massachusetts school law requires "30 cubic feet per minute per scholar;" in other words, 1,800 cubic feet per hour per capita. A theatre, or other auditorium, should have at least 1,000 cubic feet of fresh air per hour per capita, and double that quantity is not excessive. Chambers in dwelling-houses should have 1,500 cubic feet per hour per sleeper.

Even with these amounts of air moved, a room may be poorly ventilated and poorly warmed also, if proper mixing of the air is not produced within the room. This is accomplished by the positions of registers, both inlets and outlets, but it principally depends on the outlets.

The size of a room has no particular bearing on the amount of air to be admitted, if it is to be occupied continuously. Four workers or four sleepers will be about as well off in this respect in a room of 1,000 cubic feet as they would be in one of 4,000 cubic feet, provided the fresh air is admitted to both alike. If there is little or no ventilation, then the large room is the better, as the air already in it may be assumed to be pure, and it will take four times as long to vitiate it to a given standard as it will the small one.

An ordinary kerosene lamp requires the oxygen of about 40 cubic feet of air in an hour, and possibly vitiates the air as much as two persons in the same time.

Air, assumed as *unity*, is taken as the standard of weight of gases, when its temperature is 60° Fahr., and the barometer 30 inches. Air for the same weight, at a temperature of 32°, occupies 775 times the space water does, a cubic foot weighing 565 troy grains. At the temperature of 32°, 12¼ cubic feet of air weighs (very nearly) one pound avoirdupois, which increases to $13\frac{9}{10}$, $14\frac{1}{10}$, and 15, for 60, 70 and 100 degrees respectively.*

The expansion of air is nearly uniform at all temperatures, expanding about $\frac{1}{490}$ of its bulk at 32°, and for each increase of *one degree* in temperature Regnault puts it a little less, while Dr. Dalton puts it as high as $\frac{1}{483}$, and other authorities have put it at $\frac{1}{480}$; any of these ratios are near enough for small differences of temperature. The following table will show the increase or decrease *in volume* of one thousand cubic feet of air at a temperature of 32°, when the expansion is $\frac{1}{490}$.

TABLE NO. 6.

		Zero.		
Temperature ... 20°—,	10°—,	0,	10°+,	20°+,
Volume 895,	914,	935,	953,	975,
Temperature ... 32°+,	40°+,	50°+,	60°+,	
Volume 1000,	1017,	1036,	1057,	
Temperature ... 70°+,	80°+,	90°+,	100°+,	
Volume 1077.5,	1098,	1128,	1139.	

To compute the volume for other temperatures, its volume at 32° being unity, use the following—

Rule.—Divide the difference between 32° and the required temperature by 490; to the answer add *one*

---

* One pound of air at 32 degrees Fahr., under the pressure of the atmosphere (29..9 inches of mercury) will occupy a space of 12.387 cubic feet, and its specific heat is 0.2379, water being unity at the same temperature.

(whole number), if the required temperature is above 32°, but if it is below, subtract it from *one* and multiply the volume of air at 32, by it.

Example.—Find the volume a thousand cubic feet of air at 32° will have at 212°. Thus, 212° − 32° = 180° ÷ 490 = 0.367 + 1.0 = 1.367 × 1000 + 1367.0 cubic feet.

To find what a given volume of air at 70° will be at 40°.—Multiply the volume by the number corresponding to 40°, and divide by the number corresponding to 70°.

To find what a given volume at 40° will be at 70°.—Multiply by the number corresponding to 70°, and divide by the number corresponding to 40°.

Example.—Required, what a volume of 3147.0 cubic feet of air at 100° will be at 50°.—Thus, 3417 × 1036 = 3539988.0 ÷ 1139.0 = 3108.0 cubic feet.

The following table is copied from a text-book, and given as Dr. Dalton's No. 7, though it does not agree accurately with that which is given as his difference of expansion; it agrees very nearly with other tables which are given as his. It shows the increase of bulk from 75° to 680° when the volume at 32° is 1,000.

TABLE NO. 7.

Fahr.		Bulk.	Fahr.		Bulk.
Temp.	75	1099	Temp.	97	1146
"	76 Summer heat	1101	"	98	1148
"	77	1104	"	99	1150
"	78	1106	"	100	1152
"	79	1108	"	110	1173
"	80	1110	"	120	1194
"	81	1112	"	130	1215
"	82	1114	"	140	1233
"	83	1116	"	150	1255
"	84	1118	"	160	1275
"	85	1121	"	170	1295
"	86	1123	"	180	1315
"	87	1125	"	190	1334
"	88	1128	"	200	1364
"	89	1130	"	210	1372
"	90	1132	"	212 Water boils	1375
"	91	1134	"	302	1558
"	92	1136	"	392	1739
"	93	1138	"	482	1919
"	94	1140	"	572	2098
"	95	1142	"	680	2312
"	96	1144			

Air is capable of holding a certain quantity of vapor of water, or any other condensable vapor, in solution, so to speak, the proportions depending on the temperature of the air. The warmer the air is, the larger quantity it will hold, and as it becomes cool again the vapor is deposited or forms clouds or fog, which condense on anything colder than the air, leaving the air when warmed capable of taking up more moisture, to be again deposited in dew or rain. It is this property of air which gives it its drying qualities.

The atmosphere is seldom laden with moisture to its utmost, and is usually capable of taking up more moisture; the difference between the total amount of moisture the air can hold and the actual amount in it is the drying power of the air.

An absolutely dry atmosphere is hardly possible. The coldest air contains some moisture, but it is not always possible to tell how much, as air is seldom saturated to its maximum; so to find the quantity of water air at a certain temperature is capable of taking up, a quantity of the air must be cooled until the moisture becomes apparent—forming a *dew point*—when a knowledge of the quantity of moisture already in the air can be had from tables (the result of experiments of Dr. Dalton and others, who have made a study of the hygrometric state of the atmosphere) which give the greatest quantity of vapor the air is capable of containing, for the different temperatures. Thus, if air is cooled from 70 to 50, and shows condensation at the latter point, all the moisture the air is capable of taking up for 70 is the difference between the quantities of vapor at those temperatures in the table.

The object in introducing this subject and in giving the following table of the quantities of vapor air is capable of taking up, is to show the great economy there is in time and the saving in heat by having the highest possible heat in a drying room that will not injure the goods or materials to be dried.

DIAGRAM SHOWING GRAINS OF VAPOR AIR IS CAPABLE OF TAKING UP, PER CUBIC FOOT, AT DIFFERENT TEMPERATURES.

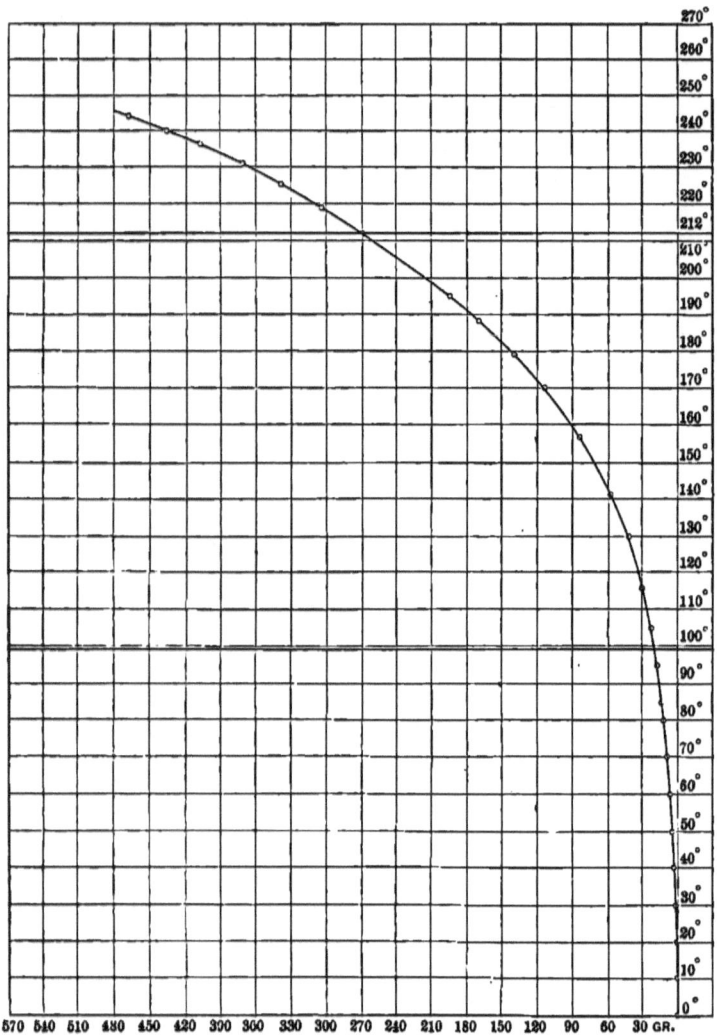

Diagram Fig. 81*b*.   *To face page* 195.

TABLE NO. 8.

A TABLE OF THE QUANTITY OF VAPOR OF WATER WHICH AIR IS CAPABLE OF ABSORBING TO THE POINT OF MAXIMUM SATURATION, IN *grains* PER CUBIC FOOT FOR VARIOUS TEMPERATURES.

Degrees Fahr.	Grains in a cubic foot.	Degrees Fahr.	Grains in a cubic foot.
10	1·1	85	12·43
15	1·31	90	14·38
20	1·56	95	16·60
25	1·85	100*	19·12
30	2·19	105	22·0
32	2·35	110	25·5
35	2·59	115	30·0
40	3·06	130	42·5
45	3·61	141	58·0
50	4·24	157	85·0
55	4·97	170	112·5
60	5·82	179	138·0
65	6·81	188	166·0
70	7·94	195	194·0
75	9·24	212	265·0
80	10·73		

It will be seen by a study of the table, that the quantity of vapor per cubic foot of air increases very rapidly as the temperature advances—*a common difference* of about 25 degrees in the rise in temperature of the air, doubling the quantity of moisture it is able to take up. Hence, all other things being equal, an increase in temperature of 25 degrees in a drying-room will reduce the time for drying about one half, and an increase of 50 degrees will reduce the time to one-fourth, and so on in that geometrical ratio.

The diagram is made to correspond to the table

* Up to 100 degrees the table has been copied from the Encyclopædia Britannica, where the full table, advancing by single degrees, can be found. Beyond 100 degrees the table has been calculated (by the author) from the elastic force of vapors according to Regnault, and is approximately correct.

No. 8, and the ordinates of the curve show grains of vapor each cubic foot of air is capable of taking up. It also shows that the quantity of moisture air is capable of taking up, agrees with the elastic force of the vapor of water. In other words, the maximum amount of vapor of water a cubic foot of air is capable of takng up, is the amount necessary to fill a cubic foot of space with steam at a pressure that would give a corresponding temperature.

The saving in heat by using a high temperature is not so apparent, as it takes just so much heat to vaporize a certain quantity of water, and the quantity of heat is a *constant*. But there is a saving in not having to heat the air and the moisture it contains from its initial temperature, so many times to carry off the given amount of moisture; in other words, the amount of heat necessary to evaporate the moisture will be the same for all temperatures, but the quantity of heat lost in the application is less.

A house 40 × 40 feet is warmed and ventilated on two stories. Each story is 11 feet in the clear, making 33,600 cubic feet, and it is desirable to change the air in the house once in each hour. In order to know its cost, a business man would proceed to figure in the following way: The steam-heater has told him the apparatus would convert 10 pounds of the return water to steam, at an expenditure of one pound of coal, consequently the next thing to know is, what is the equivalent of one pound of coal in the warming of air. Now it is admitted that a cubic foot of water, losing one degree of its heat, will warm 3,000 cubic feet of air one degree, and that one pound of it will warm 50 cubic feet of air one degree; but in convert ing the pound of water to steam, 1,000 heat units are absorbed, which, of course, will warm 50 cubic feet of air 1,000 degrees, or 500 cubic feet 100 degrees, or 5,000 cubic feet 10 degrees.* Thus the fact is established, that a pound of steam returned to water will warm 5,000 cubic feet of air 10 degrees. For the sake of safety, and to get the price as *high* as the poorest practice would make it, he takes only one-half the theoretical quantity of the coal and figures it at 7 pounds of water to the pound of coal. Thus we have $5000 \times 7 = 35,000$ cubic feet of air, which can be warmed 10 degrees by one pound of coal. But it appears that 10 pounds of coal have been burned per hour, a quantity sufficient to warm 35,000 cubic feet of air 100 degrees, while the air in the room has been only 70°. Whence, then, is this apparent discrepancy? Assume air outside to be 20° Fahr., and as it passes the heat registers it has a temperature of 120°, having been warmed just 100 degrees in passing through the indirect radiator; but an examination of the air, as it goes out at the ventilating register, shows its temperature to be 70°, which would suggest 50 degrees of the heat had been utilized in the rooms in maintaining the temperature, and the other 50 had escaped through the ventilator, and been lost as heat; but it has produced *ventilation*, and the movement of the air.

* The quantity of air, water or steam will warm, is figured according to the specific heat of each for the same weight. Approximately, water requires 4.2 times as much heat to warm a given weight of it any number of degrees as the same weight of air; but as air occupies 775 times the space water does, for the same weight, it will have to be multiplied by this factor (relative volumes), and by the heat.—Thus, $1 \times 775 \times 4.2 = 3255$. As air contains a little moisture, which must be warmed also, the odd 255 may be dropped, and is usually figured at 3,000.

Now, the ventilating flues aggregate 2 square feet of cross section, and the air, as it escapes, has a velocity of 5 feet per second in the middle of the flues, and which, if it were not for the friction of the sides, would pass 36,000 cubic feet in an hour. Making some allowance for friction, we will say 35,000 cubic feet of air passes in an hour, exactly the cubic contents of the part of the house, ventilated; taking one-half of all the heat with it, or what represents 5 pounds of the coal burned in the hour.

Thus the ventilation of a good home can be fairly done for $1\frac{1}{4}$ cents per hour, when coal costs 5 dollars per ton; less than $3\frac{1}{2}$ cents per 100,000 cubic feet of air moved under conditions which all preponderate against the price, the difference of temperature between the inside and outside being 50°, which is a high average.

There seems to be a *simple relation* between the amount of heat necessary to maintain the temperature in a room and the amount passed off in ventilation, no matter at what temperature the air passes the register entering the room, in indirect heating.

For instance, let air enter at 20°, and instead of raising its temperature to 120° it is raised to 95° as it passes into the room. The difference between the temperature of the room, 70°, and 95° and 120°, is as 1 and 2. Thus, if the windows, etc., cool a certain quantity of the air, from 120° to 70°, they will cool twice that quantity from 95° to 70° to maintain the same heat, and twice the quantity of air will have to pass out through the ventilator at half the greater difference to make room for the fresh supply necessary to keep up the heat. So the temperature at which air passes through the heat registers only affects the quantity of air moved, and not the total heat.

This also points to another result—namely, the less the difference between the temperature at which the air leaves the heat register and the temperature at which the room is to be maintained (so long as it proves sufficient), the more air there must be passed in a given time to keep up the required warmth, which will of necessity make the air purer.

A private house kept properly warm by indirect radiation alone, with air entering the rooms at about 100° Fahr., cannot be other than sufficiently ventilated for the number of persons who would ordinarily occupy it. The lower the temperature at which the air will pass the registers and maintain the heat of the room or building, the more assurance the occupant may have of the efficiency of the apparatus as a ventilator.

CHAPTER XIX.

HIGH PRESSURE STEAM USED EXPANSIVELY IN PIPES FOR POWER AND HEATING.

It has been customary, when speaking of steam-heating apparatus, to divide them into two kinds—called respectively high and low pressure; but these terms cannot now be accepted in their literal meaning, any more than high and low pressure would express the difference between non-condensing and condensing engines.

When steam has been let into pipes at any pressure and run arbitrarily to suit the convenience of some one who wants steam at a distance, under the supposition that steam will pass to any place where pipes can be put (as it will when certain conditions are complied with), such piping used to be called a high pressure system, which is now synonomous with "expansive system," and implies steam used expansively for heating.

The conditions alluded to are: The steam must be allowed to expand, to blow through in fact, if the pipes are not run on some system that provides for taking away the water at every low point in the piping; and the quantity of steam used in a given time must be sufficient to carry along the water of condensation which forms in the pipe during transmission.

Scattered buildings, heated from one source, must be heated expansively, if they have no basements, and are on different levels, and the condensed water must be taken care of by steam traps.

The system is usually attented with considerable waste of heat from imperfect steam-traps,. etc., and requires the constant vigilance of the engineer. It should not be used in single buildings when it is possible to install a gravity apparatus or a pump governor system.

Within the last 10 or 15 years this system of expanding steam through pipes has been used in the heating of towns and cities; but it is only the old system on a larger and grander scale, where instead of heating three or four buildings from one source hundreds are supplied with steam for engines and for heating purposes.

The magnitude of the apparatus generally prevents any attempt to take back the condensed water, which of necessity is wasted after it is cooled to its utmost practical limit; and as the water becomes the property of the consumer it can be used in the house for culinary purposes, and in the laundry, if the rust from wrought iron pipe, carried along with the water, will not discolor clothes.

In New York and in other cities a return pipe was used when the system was first introduced. It is now abandoned, as it became defective rapidly, although the steam pipe is still in use and in fairly good order. For some reason, not satisfactorily explained, the return pipe is eaten or destroyed very much faster than the steam pipe. In the supply pipe the steam is, of course, nearly pure, while the water in the return pipe has the fatty acids, etc., from the engines in all cases where the exhaust steam from the engines is carried into the heating apparatus and thence to the main return pipe in the ground. This is presumably one reason for the deterioration of the return pipes in city systems faster than the steam pipes.

With steam used entirely for heating purposes, I am of the opinion the return pipe lasts nearly as long as the steam pipe, when protected on the outside in a proper manner. Dampness and moisture is a factor in destroying any pipe from the outside, and those that are subject to heat and cold alternatively with moisture will rust out quickly.

In 1882 the New York Steam Company commenced the most stupendous undertaking ever contemplated for the production of steam, and carried it to a partial success in point of magnitude, and to a practical success from the engineer's standpoint.

The scheme embraced the establishment of some twelve or fourteen stations (named after the letters of the alphabet), to be distributed through the city, and to be erected as the requirements of the districts seem to demand and justify.

Two of these stations were built in New York City, one being "Station B," which is situated on the south side of Greenwich street, between Courtland and Dey streets, and is built on a ground space of about 75 feet by 150 feet. The other is uptown, in a residential portion of the city, and of much smaller capacity.

Fig. 82 is the ground plan of the Greenwich street

station and shows the irregular shape of the building, with the foundations for chimneys and piers. Fig. 83 is the first story plan, and may be properly said to represent each of the four floors on which the boilers are set. Fig. 84 is a section about on the line $A'\ B'$, and shows a view through one chimney on its greatest axis, with the approximate position and setting of the boilers and the floor trussing and columns. Fig. 85 is a section on the line $C'\ D'$, showing a chimney on its shortest axis; and Fig. 86 is the facade, whose principal feature is the immense opening in the brickwork (j), which extends from almost the roof to the water table, and which is fitted with portable or easily moved sections, to allow for the ingress and egress of heavy machinery.

The positions marked B, between the columns, are where the boilers are placed, with their fronts and fire doors facing on the fire room D. It is intended to have 4,000 *horse-power* to a floor, each position between columns containing 250 horse-power, making for the four floors of boilers an aggregate of 16,000 horse-power.

The boilers used are the Babcock and Wilcox water-tube type, and there may be said to be two boilers in a nest (between each set of pillars,) or to the same fire grate, the horizontal steam and water drums of each being connected by a large cross pipe, virtually making them one boiler.

The boilers are suspended and do not rest on the brickwork or fire front, and are set substantially as shown, the fire being under the front or high end of the tubes and the products of combustion passing between the tubes into a chamber under the steam and water drum, then down between the tubes to the back of the bridge-wall, thence up again and to the chimney.

The circulation of the water within these boilers is up through the tubes, where it enters the drum and flows backward and down again to the tubes, giving, it is claimed, extraordinary good results as a steam maker, and preventing, to a great extent, the formation

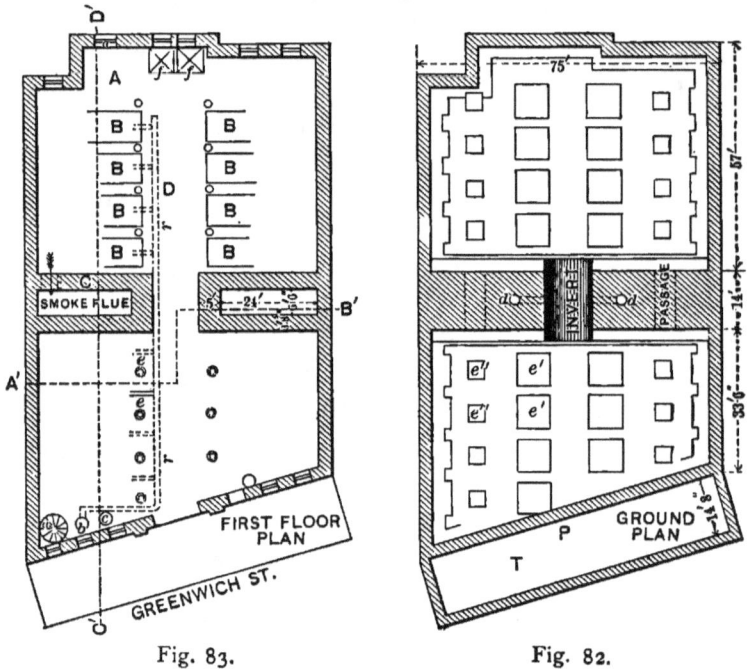

Fig. 83.   Fig. 82.

of deposits upon the inside surfaces and allowing all loose substances to settle into the mud-drum.

One of the objects for selecting these boilers for the peculiar situations they here occupy, aside from their high efficiency and the small cubic capacity they occupy for horse power, is their supposed absolute safety from any kind of burst or explosion that would materially injure the building or other boilers above or below them.

I will now follow the water from the time it enters

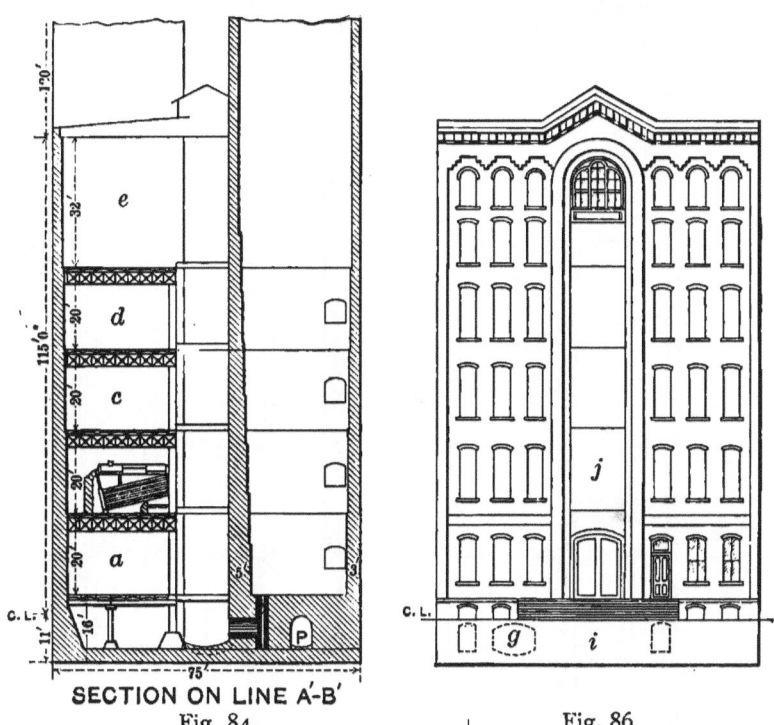

SECTION ON LINE A'-B'
Fig. 84.

Fig. 86

the boilers, through its changes and stages, to the consumer, and back again to the boilers in the station, and describe details as they suggest themselves in that order.

There are no feed water heaters used, the return water, with the steam necessary to force it around its circuit, presumably being sufficient to heat the water

206 STEAM HEATING FOR BUILDINGS.

necessary to be supplied to make up for the loss caused by steam supplied to engines.

The water enters the boilers before described, where it is converted into steam, thence it is passed through an 8-inch pipe to about an 18-inch main, see dotted lines $q\ r$, which runs above and in front of the boilers,

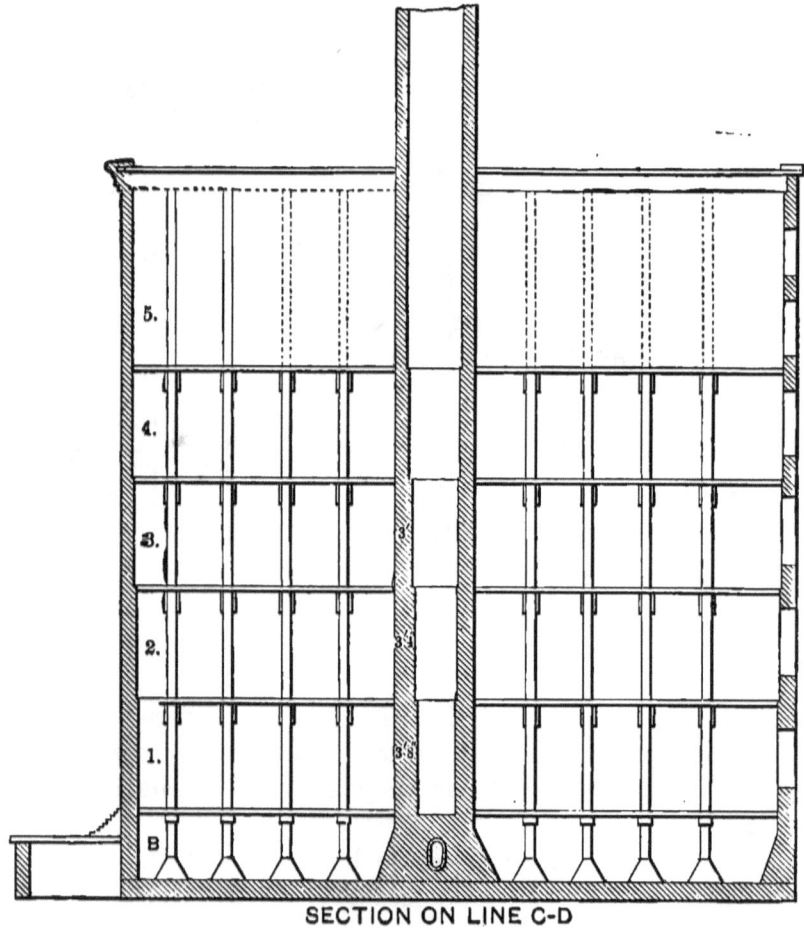

SECTION ON LINE C-D
Fig. 85.

through the fire room D, and connects with the vertical cylinder b which conveys the steam to the basement, preparatory to distributing it to the street. The main connections from the drums of the boilers to the pipe r are partly composed of copper, and are supplied with bends of long radius, to give elasticity to the branch and an easy flow to the steam. Each is fitted with an angle stop valve of special design, the position of the valve being close to the main pipe r. This valve, an idea of which can be had from the diagram, Fig. 87 is to prevent the steam from the whole system flowing backward and escaping through one boiler, should any part of it give out. It is a combined stop and check valve; the steam from the boilers as it passes into the general system having to force its way under the disk which it raises, but which would be instantly thrown down, checking a back flow from the main if the bursting of a tube caused a difference of pressure between the main and one boiler.

From the bottom of the cylinder b, which is 3′ 6″ in diameter and into which the main horizontal pipe from each floor connects, steam is distributed from the main in the street by large branches which lead off in different directions, the pipes which now lead to the street passing through the arched opening in the wall g, Fig. 86. These pipes are supplied with stop valves near the cylinder, and are for the purpose of shutting off one street section or sub-division without affecting the others.

The largest pipes laid in the street are 15 inches in diameter and run down to about 8 inches. They are wrought-iron lap welded tubes, in lengths of about

20 feet, with flanges on their ends and bolted together. The flanges are what is commonly known as "flange unions" and are faced true and brought together on a concentrically corrugated copper gasket, nothing but lead paint being used to form the joint.

The pipes are not screwed into the flanges in the ordinary way, but are inserted and expanded in the same manner as a boiler tube, a groove being made on the bore of the flange, into which the metal of the pipe is pressed by a Dudgeon expander.

From the area wall the system consists essentially

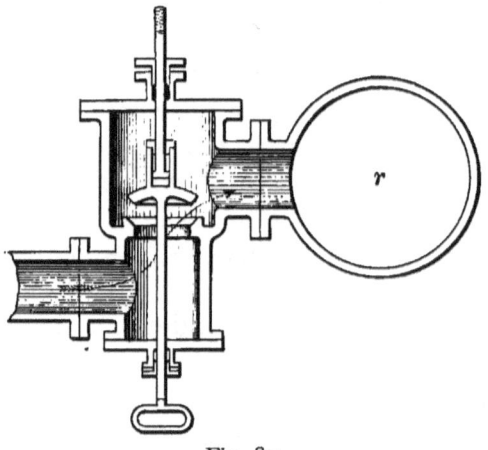

Fig. 87.

of two pipes, a larger one for conveying the steam to the consumer, and a smaller one for returning the condensed water to the station. The subdivisions of the system which radiate from the station ramify in many directions, pipes now being laid in Broadway from Warren street to Bowling Green; from Broadway down Wall street to Pearl; also down Pine and Fulton to Nassau, and Liberty street and Maiden Lane to William street; also a line in the neighborhood of Ann and Beekman street. On the other side of Broadway pipes are laid through portions of Greenwich and in Liberty, Cortlandt, Fulton, Vesey, Barclay, Park Place and Warren streets, from Broadway to Greenwich street.

These pipes are laid in a brick conduit, where they are of large diameter, and the spaces filled with "mineral wool" *i. e.*, blast furnace slag blown to a fine floss by the action of steam while hot. some of the small pipes are covered by a wooden log bored to receive them and prepared to prevent rotting.

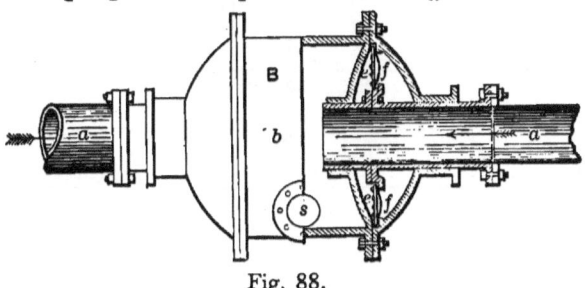

Fig. 88.

The pipes are placed at nearly uniform distance below the surface, and little attention given to inclination or pitch, simply following the grade of the ground, though avoiding a sudden dip when possible.

Bends of small magnitude, sufficient to follow the contour of the ground or slight deflections from a right-angle at street corners, made necessary by the irregular shaped crossings, are provided for by what may be called a "ball and socket" joint. It is somewhat like a flange-union, having a convex extension on one face, with a corresponding concave on the other, the faces being segments of a true circle. When put together with one of the copper gaskets before mentioned, this makes a steam-tight joint.

The water of condensation which forms within the steam pipes when everything is in full operation, is said to be comparatively small, but small as it may be, it is necessary to prevent its accumulation and not carry it long distances by forcing it ahead of the steam. At distances of about 75 feet, there is a special contrivance shown in Figures 88 and 89, which is variously called, "Expansion Joint," "Service Box,"

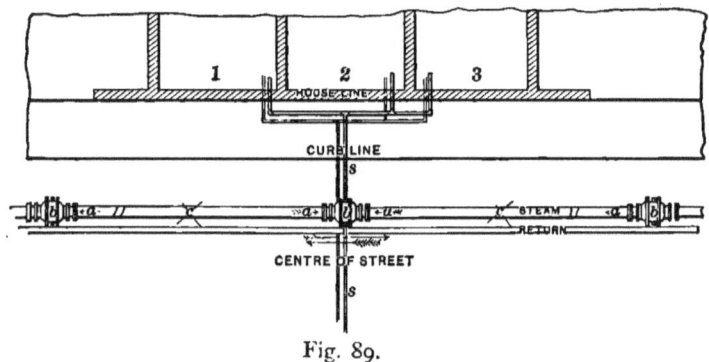

Fig. 89.

"Junction" and "Compensator," and which, in fact, does more than its name implies. Its prime object is a compensator for the expansion and contraction of the iron pipes, and it also makes a suitable fitting from which to take the house services, but in addition it is the means of freeing the mains from water. The pipes $a\ a$ are the main flow pipes. The joint itself (B) is anchored and firmly fastened, and the pipes which approach and leave it are also fastened in the middle of their lengths between it and the joints on each side of it. If now the pipe is warmed, c and c being fixed points, the movement and thrust of the pipes *a a*, must be toward *b*, which is also a fixed point, but as the pipes *a a* do not connect directly with the end of *b*, but pass inside and connect with the copper diaphragm (*e e*) the latter yields and bends inward, allowing the whole to adjust itself to the increased length. The expansion joints are numerous enough to prevent any considerable strain on the copper of the diaphragm, the intention being not to strain the metal beyond its limit of elasticity, the maximum movement when the joints are placed about 75 feet asunder being about

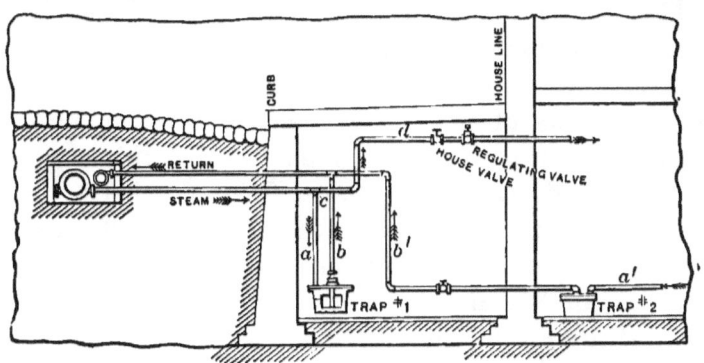

Fig. 90.

one inch. The copper diaphragms, which are concentrically corrugated, are reinforced on their outer sides by radial plates which fit closely at their adjoining edges, each forming a sector of a hollow circle, and having a rib or strengthening piece on the opposite side from the pressure. They move with the copper and are supported at their extremities in such a manner as to compensate for and accommodate themselves to the change in length; in fact, they are the real diaphragm of the apparatus, the purpose of the copper being merely to cover the joints between their edges and so render the whole steam tight.*

Condensation which forms within the pipes between any two "compensators" falls into the lower side o the annular chamber and is carried into the service-pipe (S), if the latter is in use, and if not in use is carried to the first that is in use after filling the intervening ones to the level of the bottom of the flow pipe.

The water which is thus carried out of the main through the first services which are in use, is not carried into an engine, if one is used, nor into the heating apparatus of the house, but returned again into the return water through a steam trap; this we will try to explain with the help of Fig. 90. Within the area-wall is placed a steam trap, a modification of what is known as the "Nason" principle, because the late Joseph Nason improved it and made a specialty of it, although I believe it was invented by Professor Mapes. This trap takes the water which is carried in through the house-service and falls from the point c, through the pipe a, and discharges it into the return pipe, through the pipe b, then dry steam passes through the pipe d, to the house.

Trap No. 2 is to return the water which is condensed within the house, and will drain pipes either above or below the level of the mains in the street.

This type of trap, is the only one, I believe, which can be applied to the system and take the water from

---

* Steam-fitters who are acquainted with the Handren & Robins' regulator, made by the Walworth Manufacturing Co., should readily understand this principle, as it is there used to support the copper diaphragm which operates the valve.

the cellar floor or from a sub-cellar. It is explained in the chapter on traps, and is called the "pot-trap."

The initial pressure of steam from the street is admitted as far as the "house valve," beyond which is placed the "regulating valve." The object of the latter is to cause a reduced pressure within the building. The valve used is the Curtis Regulating Valve, which can be adjusted to keep a nearly regular reduced pressure—say ten pounds on the house side, while there are 50 or 60 in the street, fluctuations in street not affecting the house pressure.

After the water is thrown by the intermittent action of the traps into the main return in the street, it flows back to the station and is received into a large tank, thence to be pumped into the boilers again. The position of the tank is in the ground under the sidewalk, as at T, Fig. 82, and a large duplex pump (P) is set close to it. This pump is kept running constantly, forcing the water through a system of feed pipes which lead to all the boilers. There is a contrivance which may be called an overflow or relief valve, arranged on the discharge pipe of the pump and loaded to a pressure of about 90 pounds, which allows the water from the pump to return to the tank, or a part of it to so return, should it not be all required at the boilers at any particular time. Each boiler is fed separately and independently, the feed valve being within the control of the fireman.

The fuel for the use of the boilers is elevated to the upper story and fed through chutes to each boiler. The boilers are fired by hand in the ordinary manner.

The ashes are carried away by a system of chutes running from the ash-pit to the basement. These chutes are furnished with a valve within the ash-pit and are within the control of the fireman. They are kept closed except to drop the ashes, otherwise the draft of the chimneys would pull on them, or should the forced draft be on, to prevent the air of the fan from escaping through them.

At the point marked A, Fig. 83, at the east end of the southeast battery of four boilers, but on the second floor, is a fan blower with a capacity of 100,000 cubic feet of air in a minute. It is driven by an engine of about 20 horse-power, and supplies air to the fires in case of necessity. It is connected with the ash-pits by a suitable system of sheet-iron ducts and takes its air supply directly from the surrounding air of the floor, but should the supply be tardy the window a can be opened.

The elevators are situated at f and f, at east end of the building, and are principally for taking up fuel to the fifth floor.

Comparatively small piping can be used in an expansion system, and when there is no provision for draining the condensed water from the pipes, a size barely sufficient to carry the required steam along is preferable; as in that case, the draft will carry the water out of the pipes; whereas, if the pipes were larger, the draft of the steam would be slow enough to cause the pipes to fill until the contracted passage increased the velocity of the steam to such a degree that it would force itself through in irregular pulsations and cause pounding.

CHAPTER XX.

EXHAUST STEAM AND ITS VALUE.

Among the many who own steam engines and the engineers who run them, there were comparatively few, until quite recently, who had a just appreciation of the thermal value of the clouds of exhaust steam continually blown to the winds from the apparently numberless exhaust pipes, which can be seen from the top of a high building in any of our large cities.

When I say that three-quarters of the practical thermal value of every pound of coal burned in the boiler furnace is lost past recovery to the consumer, I am putting it at less than the actual loss; and could this heat be converted into available motion, suitable for power purposes, it would be a boon indeed, and a fortune to the one who could do it. Perhaps there is a chance for the electrician to convert it into energy; but as yet engineers can use it for heating purposes only, where its full value can be shown in the heating of water, air, or any tangible substance.

The first purpose for which the exhaust steam is generally employed is to warm the feed water, the object being to raise its temperature as high as possible, before it enters the boiler, thereby to save fuel.

In round numbers, the warming of the feed water from 40° to 212° Fahr. can condense but $\frac{2}{11}$ of all the steam that passes through an engine; provided, of course, the quantity of water fed into the boiler is only equal to that which is required for the engine. This then leaves $\frac{9}{11}$ of all the steam that is used in engines either to be wasted or to be utilized in the heating of buildings, or in drying-rooms, cooking, water heating, or other similar purposes.

If anyone wishes to look this matter up for themselves or to convince another, it is only necessary to go to the tables and find that it requires but 172 units of heat to warm a pound of water from 40° to 212°, while a pound weight of steam at the pressure of the atmosphere contains *latent* heat equal to 967 heat units, leaving 795 heat units for some other purpose.

Among the first questions which nearly always suggest themselves to the young engineer is (1) How hot can feed water be made? (2) What percentage of the coal does the heating of the feed water represent? (3) How much of the exhaust steam from an engine can be used in heating the feed water necessary to supply the loss caused in the boiler by supplying steam to the same engine? (4) How much of it is left for use elsewhere, partly or wholly, to heat the building in winter or for drying purposes?

The answer to the first question is: Water under the pressure of the atmosphere cannot be heated above 212° Fahr., and when the feed water passes the check valve at a temperature of 200° it should be considered fairly satisfactory, although it is possible to do much better, 210° being nothing uncommon with some heaters.

Where water is forced through a heater, the temperature can be raised higher than when drawn by a pump from the heater, as the lessening of the pressure also lessens the capacity of the water for sensible heat.

Some makers of feed water heaters claim they can heat the water above 212°, because it is under pressure, but it is evidently a mistake to attempt it, as both the water to be heated and the steam necessary to heat it, would have to have a pressure above atmosphere and any attempt to keep a considerable back pressure in the exhaust pipe for the simple purpose only of warming the feed w ter above 212° is attended with a loss instead of a gain.

The attempt to heat the feed water 5° above 212° by a back pressure of 2 pounds, the mean pressure in the cylinder being 50 pounds, is attended by a loss in energy exceeding by more than five times the gain to the feed water.

The above, it must be remembered, applies only to the attempt to warm feed water above 212° Fahr. It can be made as hot as 210° Fahr. by the exhaust steam as it passes comparatively unobstructed through a good heater on its way to the outside air. On the other hand, the "no back pressure theory" must not be maintained when the exhaust steam can be all used, or even when a large percentage can be used, in warming the building. There the total gain is so very great when compared with the loss of energy in the engine that no engineer of experience will now conscientiously oppose it, and when an agent for an electric light engine or pump objects to 2 pounds back pressure on his engine on the ground that it was not designed for back pressure, reject the engine and look for one in which a reasonable back pressure above atmosphere is not objectionable.

The answer to the second question is: When the feed water is raised from *mean temperature*, 39° or 40°, to 212° by the use of the exhaust steam at atmospheric pressure, it is equivalent to very nearly two-thirteenths of the weight of the fuel necessary to convert water at *mean temperature* to steam at *any pressure*, and 15 to 18 per cent. of the coal is the greatest possible saving that can be made for this, the greatest ordinary difference of temperature.

To find the saving of other differences of temperaature in the feed water, divide the difference between the temperature of the cold water as it enters the heater and that at which it enters the boiler into 1,146, less the difference between the cold water and 32, and the product is the fraction of the coal heap.

The answer to the third question is: Two-elevenths of the exhaust steam is the greatest quantity that can be utilized in the warming of the feed water, and making a generous allowance for loss by radiation, etc., there will still be more than three-fourths of all the exhaust steam for other purposes, as was explained earlier in this chapter.

It frequently happens that an engineer, or one who sets up an engine, claims that a back pressure is injurious to the engine and reduces its efficiency or prevents its valves from working properly, and there appears to be an idea among many users of steam *that it is just as well to take live steam from the boiler* as to cause 1 or 2 pounds back pressure on the engine, the pressure necessary to get a circulation, and drive the air from all parts of the pipes and radiators.

A loss of efficiency there certainly is, but it is small and can be offset by an extra pound pressure at the boiler, and the general gain is so great that engines should be provided a little larger to meet the loss, if necessary, though as a matter of fact it is scarcely worth considering, as will be shown below.

The loss in power to an engine from back pressure is very nearly directly as the difference between back pressure and mean pressure. Thus, in an engine of 50 pounds mean pressure, with a back pressure of 2 pounds, there is a loss of 4 per cent. to the engine, and as the available energy of an engine, together with the steam used in the heater, cannot represent one-quarter of the *practical thermal* value of the coal, the loss caused by 2 pounds back pressure cannot represent 1 per cent. of the coal, and as it is an incontrovertible fact before shown, that the exhaust steam contains more than three-fourths, or 75 per cent. of the *practical* thermal value of the coal, the balance is immensely in favor of *using the exhaust steam.*

I want to guard against an error that may arise from the foregoing, and which once came under my notice. A back pressure of about five pounds was kept on an engine of about 100 horse-power for the purpose of warming one radiator of about 40 square feet of surface that was in the office of the establishment. This, of course, was poor economy, as the radiator could condense only about 10 or 12 pounds of water per hour—say $\frac{1}{4}$ to $\frac{1}{8}$ of a horse-power—while the drawback to the engine was probably 10 horse-power.

The exhaust steam from the engines of the present

day varies from 20 to 45 pounds weight per horse-power per hour, according to their class, the electric light engine of 50 to 100 horse-power using about 45 pounds of steam, and the Corliss or compound non-condensing engines doing the same work for about half that weight of steam, while ordinary commercial radiators condense from $\frac{1}{4}$ to $\frac{3}{10}$ of a pound weight of exhaust steam per square foot of surface per hour. One radiator, therefore, will cause as much or nearly as much back pressure as 100 or 500, and of course, unless the gain is greater than the loss, we do not want such an apparatus. Remember the loss is always about 2 per cent. of the engine power, which is a constant, while the gain is variable, depending on the amount of heating surface.

Let us take an example. Suppose one engine of 100 horse-power with 300 square feet of surface, and another engine of the same power with 3,000 square feet. The engine uses 40 pounds of water or steam to the horse-power, and the drawback or loss due to the back pressure is 4 per cent., which in this case is 4 horse-power, and to make up which we have to evaporate 40×4, or 160 pounds more water from the temperature of the feed water, say the equivalent of 160,000 heat units. Now, in the case of the 300 square feet of heating surface we have $300 \times .25$, or 75 pounds of water or steam condensed, the equivalent of 75,000 heat units, the gain being 85,000 heat units less than the loss; while with 3,000 square feet of radiation we have $3,000 \times .25$, or 750 pounds of water, representing 750,009 *heat units*, the gain being nearly *five* times greater than the loss. These figures are close approximations to the **facts, and any educated** engineer can work the problem out for himself when actual conditions are known; but before leaving the subject it is well to add that the total exhaust steam from a 100 horse power common engine doing full duty will warm 10,000 to 12,000 square feet of ordinary radiation. Of course if the engine is only developing half its power, the heating surface warmed will be in the same proportion, and so on.

When using the exhaust steam for the warming of the feed water alone, it is not necessary to use a back pressure valve, as the steam can be made to pass directly through the feed water heater on its way to the roof or to a condenser. It is different, however, when the exhaust steam is to be used in warming a building. In such instances sufficient back pressure must be kept in the apparatus to force the steam into the different parts of the apparatus of the building. When an apparatus is well and properly piped, 2 pounds back pressure is always sufficient for all work of ordinary magnitude. Of course, we often find an apparatus running nicely with a back pressure so low that the ordinary gauge will scarcely respond to it. This, however, is generally after the air is expelled from the apparatus and everything warmed up, in which case I have known the apparatus to run with a pressure somewhat below the atmosphere. It is well to remark here that back pressure, as the steam fitter knows it, is always pressure above atmospheric pressure. Every non-condensing engine has to exhaust against the pressure of the atmosphere and the resistance of its own exhaust pipe. When thus exhausting, they actually have a back pressure of the atmosphere, about 15 pounds, and when exhausting into a heating apparatus the total back pressure is 16 to 17 pounds. It is the pressure above the atmosphere, however, that is here called back pressure.

A back pressure valve—the form of which can be found in any steam trade catalogue—has to be used when exhaust steam is to be confined and forced into the pipes of a heating apparatus. There is a means of closing it and of loading it to open at any required pressure, and for summer use, or when steam is allowed to pass freely into the air, it can be set open, so the steam will escape freely.

When using back pressure valves, care should be taken in their selection. When used in the basement of a building, a noiseless back pressure valve should be employed, and even when used on roof the noise is telegraphed down the pipe and through the pipes of the apparatus.

Buildings are very successfully warmed by steam expanded from a high to a low pressure, through a regulating valve near the boiler. Into this low system the engines and pumps are allowed to exhaust through a suitable connection. Should the quantity of steam from the engines, etc., be greater than the coils can condense, and raise the pressure slightly, the regulating valve at the boiler will close and admit no more live steam, and should the pressure still continue to increase by the addition of exhaust steam, the back-pressure valve at the engine will open and let the excess escape to the roof through the summer exhaust pipe. This subject will be treated more fully in the succeeding chapter.

CHAPTER XXI.

EXHAUST STEAM HEATING.

When the term "exhaust steam heating" is used, it implies that the exhaust steam from the engines and pumps of a building is admitted to the heating pipes, and "an exhaust steam system of heating" implies tthat the pipes of the apparatus are arranged, or are o be arranged, for the free use of exhaust steam.

It does not, however, imply a particular method of running pipes. Any of the usual methods of piping used for low pressure steam will do, so far as the running the pipes through a building is concerned; provided, of course, the pipes are sufficiently large in diameter. It does, however, imply that proper connections will, or are to be made between the exhaust pipes of the engines, etc., and the pipes of the heating apparatus, and that proper provision will be made for taking care of the condensed water.

When the exhaust steam from the engines, etc., is not sufficient to warm the building, but is still of too much importance to be allowed to go to waste, it can be turned into the heating pipes of the house or building and condensed therein, provided the pressure carried in the heating pipes is not too great.

Should the boiler pressure be carried in the pipes, of course the exhaust steam cannot be turned into them. For this reason, therefore, in such apparatus, the boiler pressure is reduced by passing it through a pressure-reducing valve and thence allowing it to pass into the heating pipes at any pressure from $\frac{1}{2}$ pound above atmosphere upward.

Under such circumstances we will have two kinds of steam in the heating apparatus and pipes, live low pressure steam from the boilers and exhaust low pressure steam from the engines. As the exhaust steam from the engines or pumps varies in quantity as the work they do varies, the low-pressure steam from the boilers is made to respond automatically and supply the loss, and thereby keep a constant supply of mixed steam at a constant pressure in the heating apparatus to supply its demands, whether they are constant or intermittent. The large office buildings of New York and the large cities are now nearly all done on this principle, and I will, with the aid of the diagram, Fig. 91, endeavor to show the simplest form of such an apparatus, and amplify as I proceed, taking up and explaining the necessary functions and uses of the different parts of the apparatus as they are reached.

The boilers *a a* are shown at the left in perspective, with the style of connections usually employed with horizontal shell boilers. One set of the connections, *b b*, connect in o a cross main *c*—usually of considerable diameter—from one or both ends of which live high-pressure steam is taken to the engines or pumps, or any place where high pressure steam is required. In this case I show the pipe *d*, connecting

EXHAUST STEAM HEATING. 225

with, say, an electric light engine, the exhaust pipe, *e*, of which, runs first to a feed-water heater, through which it can be made to pass directly by the pipes and valves *g* and *g*, or it may go forward to the roof

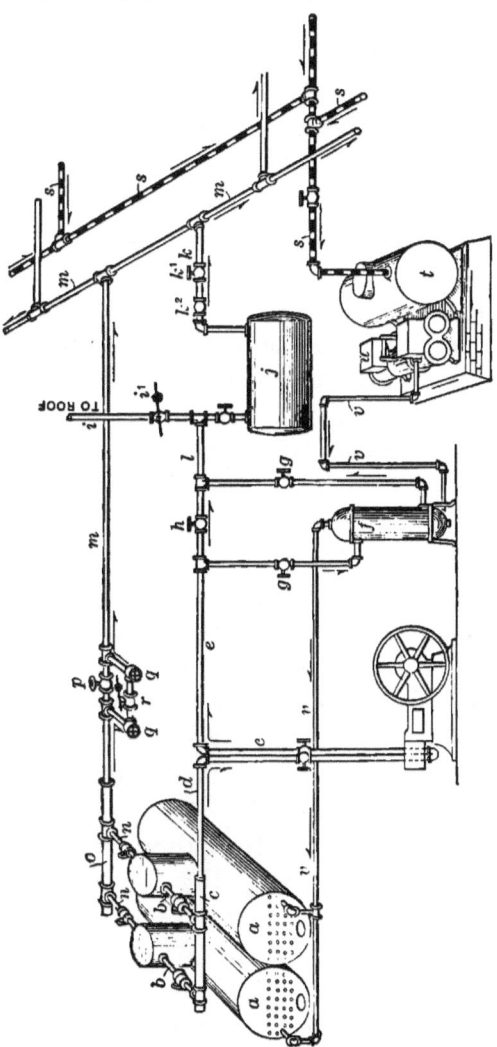

Fig. 91.

pipe i, or to the grease separator j, and the house-pipe k, by the by-pass and valve h, without the resistance of the feed water heater.

I will digress here to say a few words about feed water heater connections, as they are usually a part of an exhaust steam heating apparatus. It is always desirable to pass the exhaust steam either to the roof or to a heating apparatus with the minimum of resistance, in other words, back-pressure. Approximately only one-fifth of the exhaust steam can be utilized in the feed water heater. If four-fifths of the steam is allowed to pass through the by-pass and one-fifth through the heater by the manipulation of the valves g, g and h, this is accomplished without detriment to the temperature of the feed water and with the saving of a very considerable amount of resistance, as the tubes of a heater offer a resistance to the passage of the whole quantity of the steam that can hardly be appreciated, therefore the by-pass h should be provided, through which the four-fifths of the steam is allowed to pass freely on its way.

From the pipe l, therefore, in the diagram, the steam can pass to the exhaust pipe i, and the back pressure valve i to the roof, or it may be allowed to pass into the grease separator j, as shown. In the summer time the back-pressure valve is held open, usually by simply changing the weight to the opposite end of a lever, and the exhaust steam is allowed to pass freely to the roof, the valve on the grease separator being shut. In the winter time the back pressure valve is loaded down to the desired resistance and the steam is allowed to pass into the grease separator j on its way to the heating apparatus.

Should the steam used by the heating apparatus be less than that passed through the engine or engines, the surplus will pass off by the back-pressure valve i and the exhaust pipe, the back-pressure valve acting as a safety or escape valve.

It sometimes happens that the exhaust steam is less than the heating apparatus requires, even in moderate weather, in which case a gate valve can be put into the exhaust pipe i, just below the back-pressure valve. This is an assurance that all the exhaust steam passes into the heating mains and that none of the live steam can escape by way of the heating mains, grease separator and exhaust pipe.

The exhaust steam as it passes through an engine takes up the cylinder oil and carries it forward in volatile and finely divided particles. It is absolutely necessary that this oil be separated from the steam before it passes into the heating pipes, and later on the subject will be treated of more fully. It is sufficient for our purpose here to show its usual position j.

From this grease separator or tank then exhast steam passes to the heating pipes, more or less freed from its grease. It passes out through the pipe k, with its stop valve k^1, and sometimes with a check valve k^2.

The check valve is my own arrangement, and I have never met with it except in my own work or work for which I drew the specification. The check valve, when one is used, must be noiseless and of kind to open with small resistance. Its object is to prevent live steam from passing backwards from the heating mains m through the grease tank, and thence through the exhaust pipe to the roof, an occurence of considerable frequency, caused by carrying a too great pressure in the heating mains, when the latter is supplied by direct steam through the reducing pressure valve, or by the reducing pressure valve becoming temporarily disordered. When the check valve is used with a temporary disorder of the reducing valve, or the reducing valve being set at a higher pressure than the back-pressure valve, the excess of steam in the heating apparatus cannot escape, but withal the back-pressure will not be increased on the engine, as its exhaust steam will be free to escape through the back-pressure valve and the free exhaust pipe.

The pipe represented by *m m* is the main steam-heating pipe, from which all branches of the heating apparatus ramify, and from this point I will return to the domes of the boilers and trace the live steam from the boilers to the heating apparatus.

The pipes *n n* are dome connections, with their valves. They connect with a cross-main *o* when there is more than one boiler. From this cross-main is usually run a connection of sufficient diameter to easily supply the heating apparatus with steam, even at very low pressures, to be used for night heating or when the boilers are run for heating alone and the engines stopped.

The water of condensation from an apparatus of the class I am describing is usually returned to the boilers by an automatic pump or its equivalent, therefore in practice, when the boilers are used for heating alone, a pressure is carried suitable for running the pump, say 10 pounds. This, then, would be the lowest ordinary pressure used in the boilers, and when it is desirable to pass this steam into the heating mains at full pressure the direct pipe and valve p is opened.

When there is a high pressure in the boilers, however, say 80 or 90 pounds or more, and the engines are running, it becomes necessary to admit only sufficient live steam to make up any deficiency of exhaust, or to supply live, should the engine be stopped for the noon hour or for other reasons. In a case of this kind the direct pipe valve p is closed and the two valves $q\,q$ are opened. The pipe and valves $q\,q$, form the reducing pressure connection in combination with the reducing-pressure valve r. The pipe starts from a tee on the boiler side of the main heating valve p, and returns into the main steam-heating pipe m at any suitable position in the house side of the valve p.

This connection is usually of a smaller diameter than the main pipe. If the main is 6 inches it may be only 4 inches or 3 inches, half the diameter being fair practice. It is necessary, however, to get a size and make of valve that will not sing, and the length of pipe from the reducing valve to the main steam pipe should be as short as possible, or if it cannot be made short, it should be increased in diameter. The reason for this is, should the pipe from the regulating valve be of small diameter and long, its own resistance when passing much steam will create a fictitious pressure at the valve, which resistance will vary as the demand becomes greater or less, causing a variation of pressure at the low pressure end of the regulator, and creating the impression that the valve is not constant.

There are pressure-reducing valves in the market

that are remarkably accurate, the pressure of the steam on the low pressure side of the valve controlling the valve and keeping the pressure constant, no matter what the range of pressures may be in the boiler.

Valves q q are placed at each side of the regulating valve. This is so the high pressure from the boiler and also the low pressure from the heating mains may be shut out of the valve, that it may be adjusted or repaired. As above described and set, the valve will go on supplying steam through the pipe m to the remainder of the apparatus when it requires it, and closing automatically and retaining its steam when the supply through the exhaust connection is more than the heating apparatus can condense.

This subject would not be complete unless the manner of disposing of the condensed water was at least briefly described.

As I said before, any of the low pressure systems of piping will do for an exhaust system. Where there is a water-line system, with all the return pipes carried below a water-line, the grease collects in the vertical pipes at the water-line, provided grease is allowed to remain in the steam; and the general assumption can nearly always be that a little grease will always get through the grease separator. Of course the frequent blowing out of the system will remove this grease or oil; still I am of the opinion that the small separate return system of piping should not be selected for an exhaust steam apparatus when a large single return pipe, even if it is trapped by a water-line, can be selected. Still I would not advise the tearing out and altering of a system of small separate returns, that is otherwise in good order, should one be called upon to alter it into an exhaust system.

When the receiving tank or pump governor receptacle can be placed sufficiently low, it will remove the water-line entirely, and thus do away with the difficulty, though in most cases there is not sufficient height in basements or cellars to place the tank so low that the water will drain into it, instead of overflowing into it, as we usually find it.

In the diagram, Fig. 91, the pipes $s\ s\ s$ are the return pipes. They are further distinguished by a broken line through their center. They are shown overhead or in close position to the steam (flow) pipes $m\ m$. This, however, is not material, and they may be on the floor and overflow into the tank or receiver of the pump governor t.

The tank or receiver is provided with a float, which controls the supply of steam to the pump u. The methods used are various and cannot be shown here, but the principle is simple. Usually a water-gauge glass is used on the side of the receptacle t to indicate the level of the water, which is kept near the middle. When the water runs in through the pipe s, the float is elevated, and this in turn operates the throttle-valve of the pump, which opens and allows the pump to run. When water comes down rapidly, the pump runs all the faster, as the float is held higher. When it comes down slowly the float sinks, and the pump "creeps," as it is often called. This keeps the pump hot and always ready for immediate rapid action when an intermittent flow of water comes, which often seems to be the rule.

Pumps for this work should not be too large.

About 60 strokes per minute, when the maximum work is being done, is good practice. Duplex or some order of pumps that are always in action without attention should be used.

From the pump the water is forced into the boilers in the usual manner, and generally through the feed water heater. The pipe v v shows its usual course. The steam and exhaust pipes of the pump are omitted, but they are similar to those of any other engine. They start from the pipe c or d, and return to the exhaust pipe e, if possible, at a point before the latter pipe reaches the feed water heater.

There are other arrangements of the apparatus than those I show in the diagram. All the essential features, however, are covered. The modifications often include the placing of the grease separator before the feed water heater. This prevents the grease from going into the feed water heater in considerable quantities. However, it brings the grease separator into use the whole year, summer as well as winter. There is no objection to this if the grease separator is well covered, and there may be an advantage in keeping the grease out of the heater. I generally arrange my own form of grease separator, which is a comparatively large tank, so that the exhaust steam passes through it at all times. As it is a large receptacle, the engines exhaust into it freely, and it acts as a cushion, taking the pulsations of the engines off the heating pipes and allowing a constant flow of the exhaust steam into the system. Under the head of grease separation I will show methods of connections, etc.

CHAPTER XXII.

THE SEPARATION OF GREASE FROM EXHAUST STEAM.

Where the exhaust steam is used for heating and the water of condensation from the heating system is returned to the boiler, it is found to carry along with it a portion of the oil which has been used in steam cylinders of the engines as a lubricant. Opinions do not now vary as to the effect on the boiler, though the result in different cases depends upon the quantity and quality of the lubricant getting into the boilers, the quality of the water and the chemical composition of the various dissolvents or "scalers" employed to free the interior of the boiler from sediment. These combinations are so varied, and the trouble from this source is so dangerous and expensive, that a safe rule is to avoid the main cause of danger, the introduction of oil or grease, to the boilers.

The boilers in the Rogers building of the Massachusetts Institute of Technology in Boston were destroyed a few years ago by grease that was carried into them from the neighboring new building of the group. The contractor used the exhaust steam and provided no means of separating the grease from it, and until the injury was done, no one was aware of the mischief that was going on. They now use a grease separator of the kind first used and designed by myself.

In New York we have had several grievous cases of the same kind in office buildings and hotels where improper or no grease separators were used, and much damage is being done all the time by improper apparatus. Small apparatus of the "turkey gizzard" order will not take sufficient of the oil and grease of the steam. They may remove 80 or 90 per cent. of the oil, but the remaining 10 to 20 per cent. is sufficient to cause the burning of the boiler, particularly if it is a shell boiler wherein the heavy "slugs" of grease can gravitate to the hottest part of the shell of the boiler, the bottom, and become attached thereto.

An essential feature of any grease separator to be be effective is size. Small convoluted or corrugated bulbs or globes will not remove sufficient grease.

The steam passes over their rubbing surfaces with too much velocity and carries the particles of grease forward with it, almost with the same ease that it carries them through the exhaust pipe of the engine.

The only method I have found to be certain in grease separation is to project the steam on the surface of a body of water in a large tank. The water may be comparatively hot, as hot, in fact, as the steam can make it, still it will hold the oil, if it once reaches its surface. As I said before, however, size, is an important element. When a small confined apparatus is used, the velocity of the steam through any part of it is comparatively little less than through a section of exhaust pipe itself.

When the steam has a rapid motion it forces the particles of grease along with it, holding many of them in suspension and forcing those that have already been deposited by contact with the sides or ribs of the apparatus along with it also. It thus carries it forward into the pipe again and thence into the heating apparatus; whence it is carried to the boiler with the "return" or condensed water.

This makes two essential points, therefore, necessary for a thorough grease separator, when grease has to be separated from steam. The fact is to project the steam and grease against the surface of the water, and the second is to slow down the velocity of the

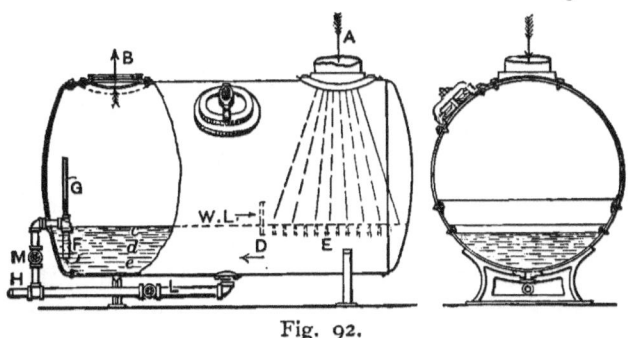

Fig. 92.

steam as it passes on its way to the outlet. The first is easy enough, but at first thought, the second may not be so apparent. It is done, however, by enlarging the receptical through which the steam passes. When the use of exhaust steam in New York buildings began to become general, the problem was presented to me, and I conceived the idea of a tank of comparatively large size, at the bottom of which was held a quantity of water of about one-third the capacity of the tank. Into this at one end the steam was made to enter at the top, so as to throw the particles of oil and water held in suspension directly downwards and against the surface of the water. It was necessary, however, to prevent too great a commotion in the tank and to slow down the velocity of the steam as it passed away from the surface of the water. So to do this the tank was enlarged in diameter until the steam in passing through it, was comparatively quiet. I use tanks arranged as shown in Fig. 92.

I find that an 8-inch exhaust pipe to a 48-inch tank placed on its side, is about the safe limit of proportions. The tank's length should be about one and a half times its diameter and the flanges as near the ends as possible.

When room is a considerable object, I use a tank on its end as shown in Fig. 93. The distance from the inlet to the outlet is less than with the horizontal form, but the area of the tank can be increased by its length, which is only limited by the height of the basement or cellar of the building.

With tanks like the above there is no difficulty in reducing the velocity of the steam to one-thirtieth or one-fiftieth of what it has in the pipe, and under such conditions the projection of the particles of the grease against the water appear to be unaffected by any tendency of the steam to find a short cut to the outlet.

Another advantage such an apparatus has, is that it takes the violent pulsations of he engine off the back pressure valve and aids materially to stop the noises that are communicated to the walls and floors of a building by the quick running electric engines. The illustrations Figs. 92 and 93 have the same reference letters. A is the exhaust inlet from the engine; B the exhaust outlet to the outside or the heating system; c, the lighter oils floating upon clean, hot water; d, hot water; e, sediment; D, a dam to prevent the oil returning and overflowing the bars E; E, bars for the same purpose; F, inverted overflow pipe and water level; G, anti-syphon pipe; H, pipe to the blow-off-tank or sewer; L, blow-off valve: M, overflow valve; I, back pressure valve; J, check valve controlling heating system. The body of the separator is made of about $\frac{1}{4}$ inch iron.

The operation of the apparatus is as follows: The exhaust steam from the engine enters the separator through the pipe A, and is projected directly on the surface of the water already contained in the tank, the oil and water settling at their separate and natural levels, and the steam turning upward and outward through the pipe B to the external air or the heating system. When the water has reached the overflow level, it will continue passing up through the inverted overflow pipe F, which always draws its supply from the lower end, being sufficiently submerged to be below the floating oil. This keeps a constant level in the apparatus without attention on the engineer's part, excepting to draw the contents of the apparatus off now and then, once a week or a month as the case may be, through the pipe and valve L. The regular overflow through the pipe F and valve M is clean water that will not clog or effect the pipes. The action is continuous as long as the engine is running, the accumulation of water simply overflowing. The function of the pipe G is to prevent syphonage. Without it the pressure in the tank would charge the pipes F and M and syphon the entire contents away.

By the arrangement of the dam D and the cells formed by the bars E, clear water is always exposed to the greasy stream as it is projected downwards. The entrained water in the steam has already collected the grease, so that the globules of water and grease are thrown on the water in the cells between bars. The impulse of the steam depresses the level of the water in the cells and compels it to move towards the overflow end at F. The return wave of the water is at the surface, and flows in the opposite direction against the dam, over which it does not go. The continued pulsating surface of the water between the bars tends to force the oil to the opposite side of the dam, thus always presenting a fresh clean surface of water for the reception of the oil.

The exhaust steam, freed of its oil, passes out through the pipe B and valve T for use in the heating system, and when condensed can be safely returned by the feed pump for further use in the boiler. The small amount of water deposited and wasted from the separator does not repay an effort to return it to the boiler, though it may be of value for other uses, and in some cases it is used and allowed to overflow into the open tank of the hydraulic elevator system, where the greasy water is found to be a considerable factor in keeping the hydraulic cylinders in good order.

In cleaning the separator, the stop valve M is closed and the sediment valve on the bottom is opened, when the pressure will purge it.

Of course if the water from the separator is to be utilized, a separate connection to the sewer should be provided for use in cleaning.

SEPARATION OF GREASE, ETC. 239

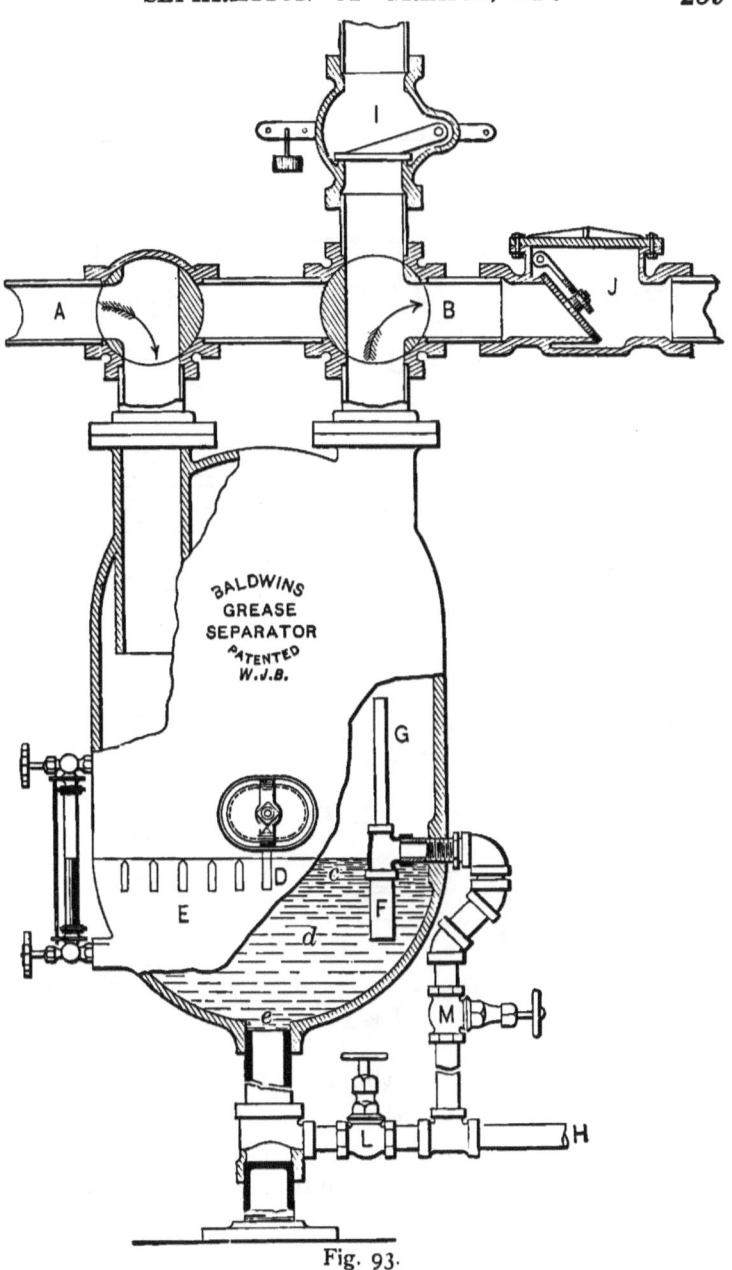

Fig. 93.

To make the apparatus entirely automatic a "Royal" steam trap can be introduced into the pipe M and allowed to waste into the pipe H. A water glass may

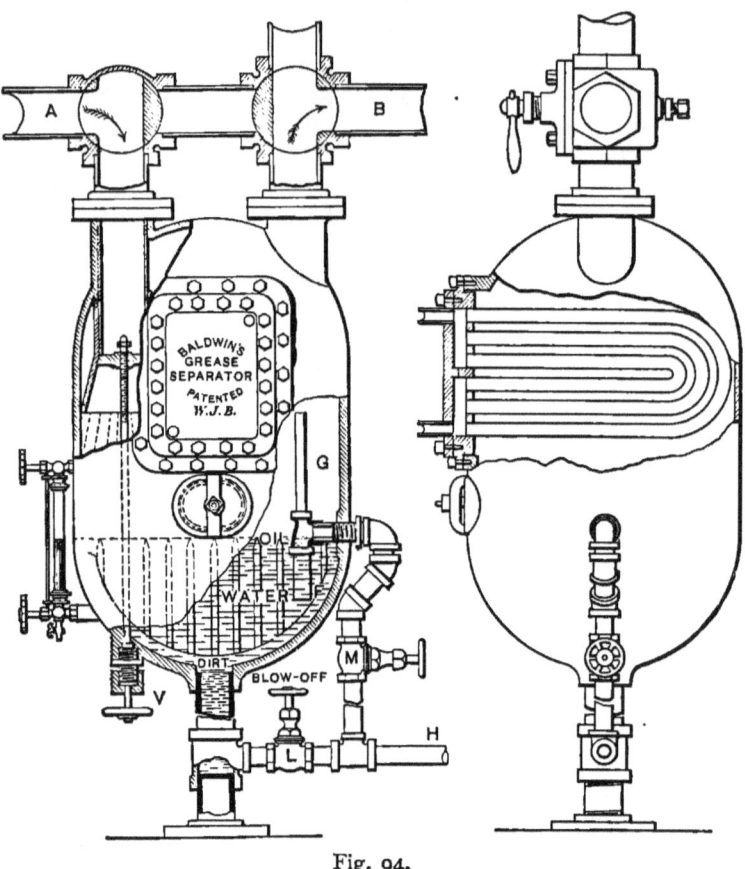

Fig. 94.

be placed on the side of the apparatus to show at a glance the level of the water.

The position of the grease separator is an important question. In the foregoing chapter I show it following the feed water heater. Some prefer to place it in the system in such a manner that the steam will pass through it before it passes through the feed water heater. There may be some advantage in this by keeping grease out of the heater. All heaters, however, play some part in the separation of grease from steam, and were it not for this fact alone, there would be a great many more burned boilers in our city buildings. Of course when the feed water heater is on a branch of the exhaust pipe it cannot act as a grease separator to any appreciable extent.

Fig. 94 shows a combined grease separator and feed water heater which I designed. The steam is first projected against the water and then allowed to pass through the feed water heater, which is suspended directly in the grease separating tank. It saves in cost and room, makes a very efficient heater and it is supposed by some to take exceedingly fine particles of oil out of the steam that may by some means escape the water. In the Manhattan and Merchants' Bank Building, 40 and 42 Wall street, New York, one of these water tank grease separators can be seen where it has been in use since 1886.

The water of condensation is collected in a receiving tank. and in all those years no grease has accumulated in the receiving tank, not to consider the boilers at all.

CHAPTER XXIII.

BOILING AND COOKING BY STEAM, AND HINTS AS TO HOW THE APPARATUS SHOULD BE CONNECTED.

LARGE institutions with many inmates find it almost impossible to cook without the aid of steam, and manufacturers have long since abandoned all externally fired kettles for this purpose. Of the superiority of steam as a means of drying and cooking, there is no question, and the occasional failures which occur should not be attributed to steam, but to errors in the construction of apparatus, and an ignorance of their use. Satisfactory appliances are within the reach of the steam-fitter, though frequently the ruinous competition in small things, which compels the lowest bidder to neglect and omit everything possible, or in other words, "to do the least for the least money," ruins the effect of otherwise successful machines.

The first and commonest kind of cooking by steam is "steaming," which is again divided into steaming in the atmosphere (or at atmospheric pressure), and steaming under pressure in closed tanks or boilers. Steaming can be used in the preparation of anything into which water cannot enter or become part of, as oils, or of substances which want an addition of water, but are capable of taking up only sufficient water to properly prepare them as vegetables, or substances which want to be bleached or disintegrated, as rags.

The simplest form of steamer is the ordinary kitchen steamer; a wire basket or tin pot with holes in the bottom of it, suspended in a larger pot with water in the bottom of the latter, the water not reaching the

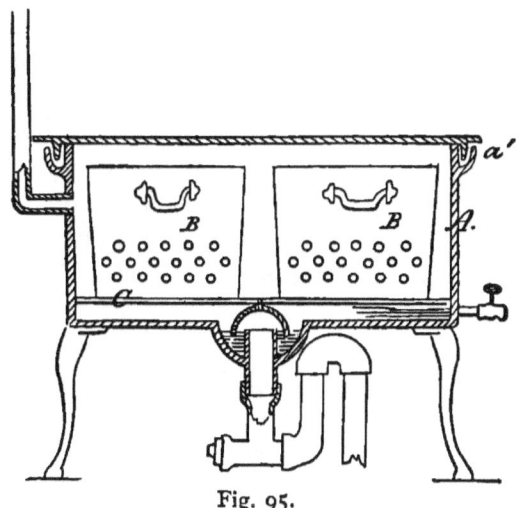

Fig. 95.

bottom of the basket. The steam, rising and mixing with the air in the basket, gives a uniform heat when the water in the lower pot is boiling.

It is well known to the intelligent cook, that vegetable-cooked this way can be done through without breaking or without losing any of their starch. This cannot be done in boiling water, as the mechanical action of the water during ebullition breaks and washes out part of the substances before they can be sufficiently cooked in the center.

The modification of this simplest kitchen steamer, used in large buildings, such as hotels and public institutions, is shown in Fig. 95.

The outside case, A, may be of cast-iron, or sheet-iron riveted and soldered with a cover of sheet-iron. The baskets, BB, rest inside the outer case on a perforated shelf, C, and are usually made of heavy tin plate, with holes in the bottom for the condensed steam to run off.

The connections to these steamers require particular attention, far more than would appear from a superficial examination. The condensed water which gathers in the bottom of the outside case should be carried to the sewer or drain, and must be connected in such a way that the foul air of the sewer cannot return into the steamer and contaminate the food. And as much —and more—attention should be paid to the waste connection from a vegetable steamer than is paid to the connections from a wash basin, even in a sleeping room. It is not only essential how the connections are run, but from what material they are composed, and further, how the joints are made, and from what material.

As the steam and hot water are capable of destroying lead pipes and traps, or working the lead joints out of cast-iron pipes, it is best to use either wrought-iron screwed pipe, or cast-iron pipe with rust joints, using a very deep S-trap, constructed of fittings, with plugs at every corner, so as to get straight openings at every part of the pipe, by simply removing the plugs. This is necessary to remove grease, or any obstruction that may pass into the pipe. The pipes should be of large diameter (about 3 inches) with the trap sufficiently deep to prevent the pressure of steam within the steamer from blowing it out, and connected with some contrivance, vacuum valve or vent pipe, run on approved sanitary principles, to prevent its siphoning out, as is common to all soil pipes.

There is another source of contamination or poison in the connections of vegetable steamers, or any other steam boiler, which must have a vapor pipe. These pipes should not be constructed of galvanized iron or copper, or any other substance whose salts are poisonous, as the condensation which takes place within this vapor pipe falls back into the kettle, continually washing down carbonate or sulphate, or whatever may be formed, that yields easily to the action of pure water. These pipes should be constructed of iron gas pipe, with screwed joints, or cast-iron pipes with rust joints.

The live steam connection to an open steam box or steamer should be very small. Usually a $\frac{3}{4}$ or $\frac{1}{2}$-inch pipe is used, and there is no discretion exercised in the manipulation, but an endeavor made to cook as rapidly as possible, regardless of steam. Beyond a certain quantity of steam admitted, nothing is gained in time, as just steam enough to expel the air is all that can be used; a greater supply is only wasted through the vapor pipe, or escapes into the kitchen under the edge of the cover.

There is another point in the construction of open steamers worth considering, namely, a water seal around the edge of the cover. The seal consists of a groove or channel around the top edge of the case, into which a rib around the under side of the cover fits, as can be seen at a', Fig. 95. This seal should be as deep as possible, and to be effective should run around the whole cover, and not be dispensed with on the side of the hinges, as is frequently done. The object of this water trap, or seal, is to prevent steam from escaping into the kitchen and to force any excess of pressure out through the vapor pipe. To get the greatest economy, the water seal should be 2 inches or more deep, with a small sized vapor pipe with a valve in it, so it can be choked down to hold a pressure in the steamer, but not enough to force the water seal.

Steaming under pressure must be done in a closed

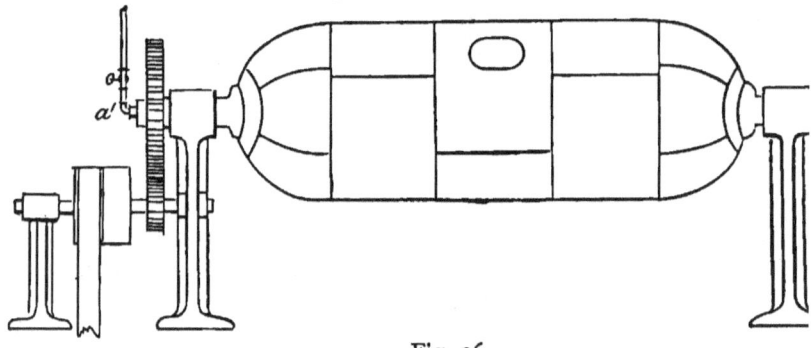

Fig. 96.

boiler or tight tank capable of resisting high pressure steam. A common form of this class of steamers is the rag boiler in the paper mill. It is a horizontal cylinder, with conical ends, supported on trunions, and made to revolve by machinery, so as to use the mechanical motion in assisting the disintegration of the rags. This boiler is shown in Fig. 96, and should be constructed of exceedingly heavy iron, or it may explode and do much injury. The pipe connections are made at the ends of the trunions, a, which are provided with stuffing-boxes revolving around the pipe, thus leaving it stationary.

Another form of high pressure steamer is an upright tank of strong construction, in which fats are rendered and separated by the action of high pressure steam. This tank is shown at Fig. 97, and is often 20 to 30 feet long. The fats and oils stratify according to their gravity, with the water of condensation underneath, and are drawn off at the numerous cocks, according to their quality.

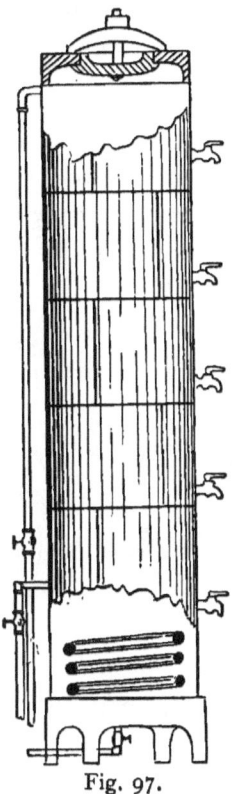

Fig. 97.

The steam connections on these tanks are made at the top and bottom, and they are sometimes constructed with a spiral coil near the bottom.

Cooking and manufacturing by the transmission of steam heat through metal surfaces and not by direct contact, as in steaming, requires apparatus of varied designs, often the result of years of experimenting, the following modifications being the most common

Figs. 98 and 99 shows sections of two of the ordinary forms of double-bottomed steam cooking kettles.

The various uses to which these kettles are applied are wonderful. Differing very little in shape, the size alone adapts them to the special use. Small sizes, 20 to 40 gallons, can be used for glue melting, etc.; sizes running from 60 to 100 gallons are mostly used in hotels and institutions for cooking meats and farinaceous foods, and larger ones, up to 500 gallons, are used in sugar-houses and soap-boiling establishments.

Sizes to 200 gallons are usually cast-iron, but larger ones are often made of wrought iron, riveted and calked.

The connections to these kettles are plain, but the

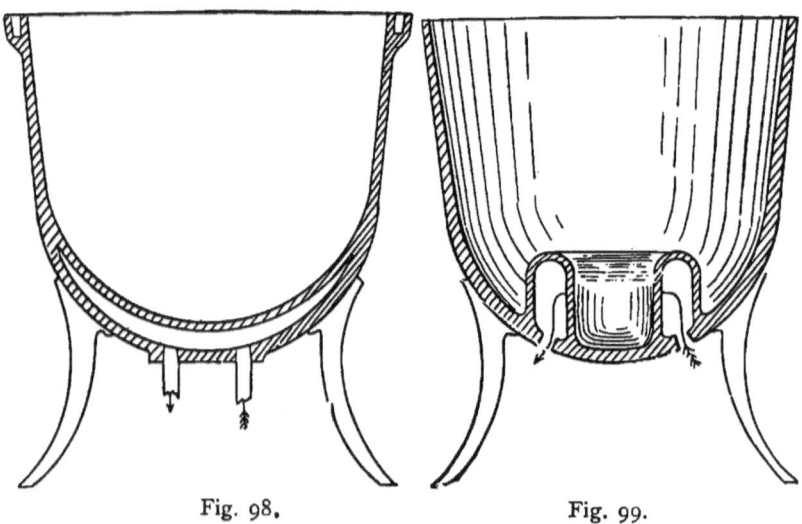

Fig. 98. Fig. 99.

steam pipe should be large, and the return water pipe should not be put back into a return gravity circulation apparatus, but should be carried away by a good steam trap of approved pattern.

Vapor pipes from these kettles should be of iron, for the same reason mentioned in connection with "steamers."

The pipe from the inside of the kettle, which carries

the contents to a receptacle or sewer, should be large, with tees and plugs at every right angle, instead of elbows, to permit of easy and rapid cleaning should it get stopped with grease or any other substance which hardens on cooling.

In these kettles steam cannot be wasted unless it is passed through a defective steam trap, the consumption of steam depending on the amount of work to be done and the radiation from its outer sides.

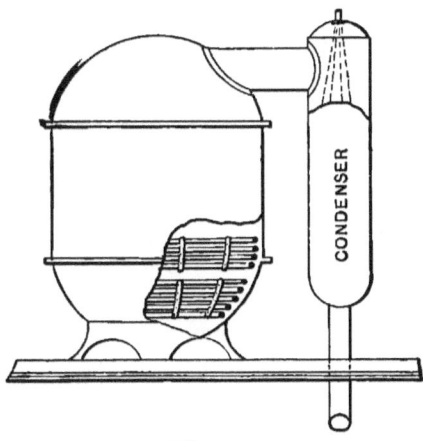

Fig. 100.

This radiation is often partly prevented by an outside loose jacket, and if the space between the jacket and the kettle is filled with some non-conducting material, the loss of heat from the outside of the kettle will be reduced to a minimum.

There is another class of kettles or pans which are not double-bottomed, but boil and cook by steam heat transmitted through spiral oils, passed around the inside of the bottom, the pan itself being partially exhausted of atmosphere in order that the contents may boil at a temperature much below 212° Fahr. These kettles or pans are usually very large, and are principally used in sugar-houses and condensed milk establishments, or any place where boiling or evaporating at very low temperatures is desired.

Fig. 100 shows a section of one of these pans. The principal point of importance to the steam-fitter or coppersmith are the sizes of the pipes and the manner in which they should be run.

When a quantity of water is to be raised from ordinary temperatures (35° to 55° Fahr.) to boiling, it must be borne in mind by the fitter, or constructor, that it will take in steam at least $\frac{1}{8}$ of the weight of the water in the pan to raise it to the boiling point, and that when steam is first turned into the space between the bottoms of jacketed kettles or into the spiral coils of tanks or vacuum pans, the shrinkage, *i. e.*, condensation of the steam, for the greater differences of temperature is something enormous, and unless the supply of steam is continuous and high and the pipe which conveys it ample for the greatest amount of work that can be put on it at any one time, the result will be the filling up of the space or coil with water.

This is a point of more importance than appears at first. It is a common thing to find a stock-boiler or a kettle of any kind that will not begin to boil for hours after steam is let on, provided the kettle has been filled with water before the steam was turned on. It is also found that should the steam be turned on first and the water run into the kettle gradually afterwards, that in most cases, by the time the kettle is full, it will also be boiling. Now the reason is plain enough upon a little consideration. When the kettle is full of cold water it is in a condition to be a good condenser, and when the steam is let into the coil or steam space, it is instantly condensed and there is a vacuum or a very small pressure within the coil or steam space. This prevents the condensed water from running off into the condensed water pipes, where there is always considerable back pressure, and thus the coil or space becomes filled with slightly warm water, while the kettle is filled with cold water. This forms a deadlock for a time. The steam, however, ceases to condense when it cannot pass into the coil or steam space, and then its pressure asserts itself and a little of the water is forced out of the coil and into the return pipe; but only a little, for just as soon as the steam meets the cold water again its pressure is reduced and a very small part of the coil only is active, the upper part, or the part near the inlet. This goes on for an hour or two or three, and then the steam gradually overcomes the cold of the water, as the condensing power of the water becomes less.

Of course if the pressure of steam is great enough and the pipes very large, the steam will assert its influence quickly. We are therefore to use as large pipes as practicable in such cases, and coils of large diameter and a few turns in preference to long coils of small diameter pipe.

With small inlet pipes and coils the trouble may be overcome by turning on the steam and admitting the water slowly, but this cannot always be practicable.

Long, flat wooden vats, with a coil of any convenient shape, in the bottom, are often used for the evaporation of the water from brine by the salt manufacturer. Exhaust steam from neighboring engines can be used here to advantage, thus utilizing heat that would otherwise be lost.

Another common way of warming or boiling water, when the object is not evaporation, but the warming of a tank of water for laundry purposes, or when the addition of the condensed steam is a benefit (provided it is not greasy), is to put the steam pipe directly into the water in the form of an open butt or a perforated coil. This mode is usually attended with noise, but it is quick and effective.

When a perforated coil is used, it is usual for the

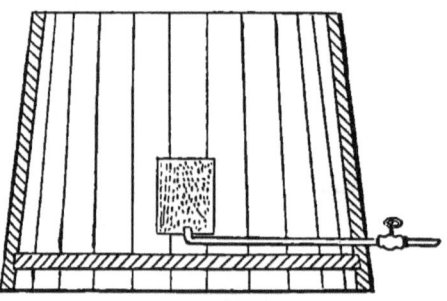

Fig. 101.

fitter to have as many small holes in the coil as will be equal in the aggregate to the area of cross-section of the pipe in the coil, but in practice this is not nearly sufficient, if he wants to pass out all, or nearly all, of the steam and water which the supply-pipe is capable of passing.

Within an empty pipe, steam has a very high velocity, but striking the water as it passes the holes retards it so much that five to ten times the area of the pipe in small holes has not been found too great in practice, the time of boiling lessening rapidly up to ten times with shallow water and forty pounds of steam.

The pressure of the steam and the depth of the water affects the time of heating; high pressure accelerates and deep water retards it.

The lower the pressure of the steam that will pass out, as it strikes the water, the less the noise will be, and a good way to avoid noises is to have a coil or pipe of large diameter in the water, with a great many small holes in it, letting the high pressure steam expand into this perforated pipe through a "throttled" valve, until the desired low pressure is attained.

Another way to prevent noise is to place an iron cylinder with wire-cloth ends, filled with shot, over the end of the steam-pipe, the pipe turned up into the cylinder, and the cylinder in a vertical position. (See Fig. 101.)

Still another way to warm water with steam is at the nozzle or cock where it is drawn. A very simple method is by mingling the steam and the water after they pass their respective cocks or valves, as shown at Fig. 102. There should be no cock or valve put in the bib, a', for closing it will either force the water or steam (whichever has the greatest pressure) into the other. Therefore it is necessary to have little or no resistance in the pipe after passing the valves.

A very simple noiseless nozzle is shown in Fig. 103 It consists of an enlargement after passing the valves filled with shot, with a strainer to prevent the shot from passing out; or it may be filled with clean gravel or anything the steam and water will have the least action on. By the regulation of the valves, a steady stream of water of almost any temperature between 212° and the temperature of the cold water can be had.

Often the pipe-fitter is called upon to construct apparatus to warm water for bath-houses, laundries, or any place where they have no steam and require no power, hence do not wish to have a steam boiler, but nevertheless use more water than can be warmed by the ordinary water-back in the stove. The problem is then, to warm the largest amount of water with the smallest expenditure of fuel. Fig. 104 shows an ap-

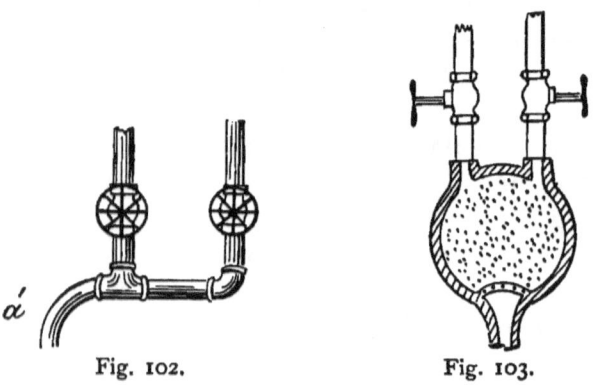

Fig. 102.   Fig. 103.

paratus that for permanency and cost of maintenacce is very satisfactory. A, is a tank of any convenient shape; B, a cast or wrought-iron boiler, similar to that used for green house heating; C, a connection from top of boiler to the side of the tank, not very high up, as all the water below the point where it enters the tank cannot be estimated as part of the working capacity of the tank, for it is necessary to always keep this pipe covered with the water; D, the return pipe from the tank to the boiler, its inner end being carried a few inches above the bottom of the tank to prevent sediment from being carried into the boiler; and *E* the pipe leading from the tank for the distribution of the hot water, the position it occupies being important, as it must always be above the pipe *C*, to prevent the possibility of drawing the water in the tank entirely down to that point.

The tank may be furnished with a ball cock to the cold-water pipe, as shown at *F*, to keep a constant level of water.

By feeding the water into the tank instead of into

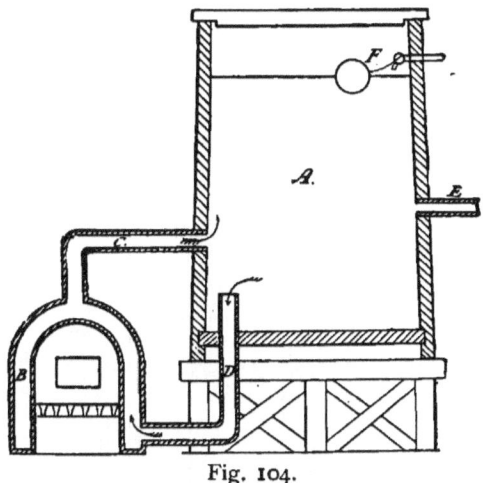

Fig. 104.

the boiler, impurities are deposited in the bottom of it instead of being carried into the boiler. The same is true of all hot water apparatus, if the bottom of the tank is below the return pipe, with capacity enough in the tank to prevent rapid currents.

The above is one method often used in warming swimming baths. A boiler of the hot water pattern is set up at some convenient part of the building, below the swimming tank, and circulating pipes *C*

256   *STEAM HEATING FOR BUILDINGS*

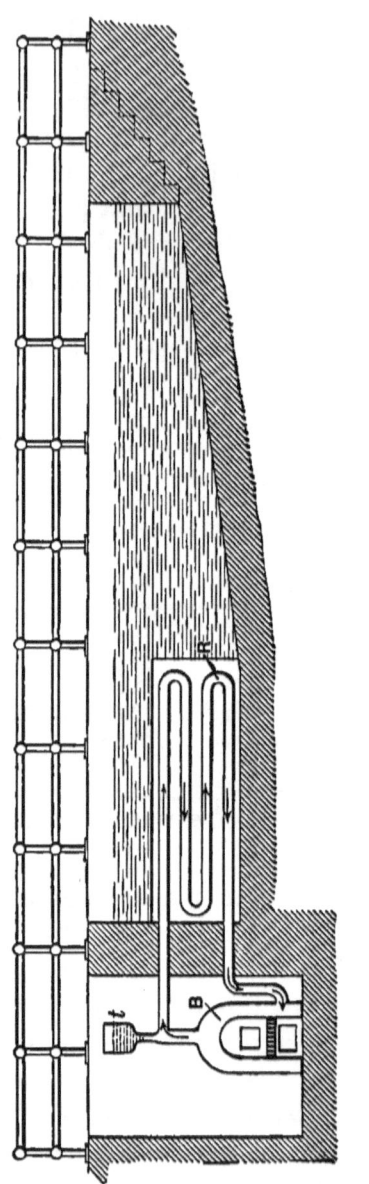

Fig. 105.

and *D* are connected with the tank. The tank may be of any desired shape and size, but otherwise the apparatus is the same as that shown in Fig. 104

Although the foregoing method is often used for warming large tanks of water, it has one objection, which becomes greater as the water contains impurities. Nearly all the water of the tank is made to pass through the boiler at some time, and in doing so it deposits mud and other substances it may hold in suspension within the boiler. To obviate this an apparatus similar to Fig. 105 may be used. *T* represents the swimming tank, on each side of which there may be one turn of large diameter pipe. The pipe should be in a recess *r* in the side wall, so there will be no likelihood of a diver striking his head. In a pit at the end of the tank is a boiler *B* with an expansion tank, and in other respects a regular hot-water apparatus. In such an apparatus the water is always the same, and of course there is no deposit of mud in the boiler.

In small apparatus for making hot water, a coil of pipe is sometimes used in a stove instead of a boiler, but it often fills with mud or lime and burns out.

Before leaving the question of boiling or cooking I will refer to Fig. 106, which shows a pair of steam roasting ovens. They are cast-iron with double bottoms and double sides for about two-thirds of their heights, the double side forming a terrace or step on the inside of the oven. It is within this space that the steam circulates. Tight-fitting heavy covers fit over one-half the top to retain the hot vapors given off by the meats. They are connected similar to an ordinary radiator, and are becoming much liked in public institutions, an oven being capable of holding 60 to 70 pounds of meat, and cooking it in an hour and one half with forty pounds of pressure of steam.

Fig. 106.

Meats cooked in one of these ovens have all the appearance of a pot roast and will become tender when the same part of the animal in the ordinary oven will be hardly fit for food.

CHAPTER XXIV.

DRYING BY DIRECT STEAM.

THREE-FOURTHS of all the manufacturers outside of the metal trades, and even many of them, use heat for drying purposes; and various as are the manufacturers, so various are the modes of drying, in many instances satisfactory results being attained only by years of experience. No manufacturer of wooden articles can get along without a drying kiln. The laundry man or woman, the dyer, the hatter, the shoemaker, the tobacconnist, the piano and organ maker, the dried-fruit manufacturer, the japanner, the tanner, all must have a means of drying faster and more conveniently than can be had by exposure out-of-doors, and even now the common red bricks are often dried by artificial heat and forced currents of air when the weather is too cold or damp to do it in the open yard.

Usually steam is used in drying rooms and drying kilns because of its cleanliness, its even distribution, its safety from fire, its easy and quick management, and the cheapness of its maintenance.

The higher the temperature of a drying room, the cheaper can the articles be dried. This may not appear plain at first to those who have studied the laws of equivalents, but nevertheless it is so, being caused by local conditions, which always prevent the utiliza-

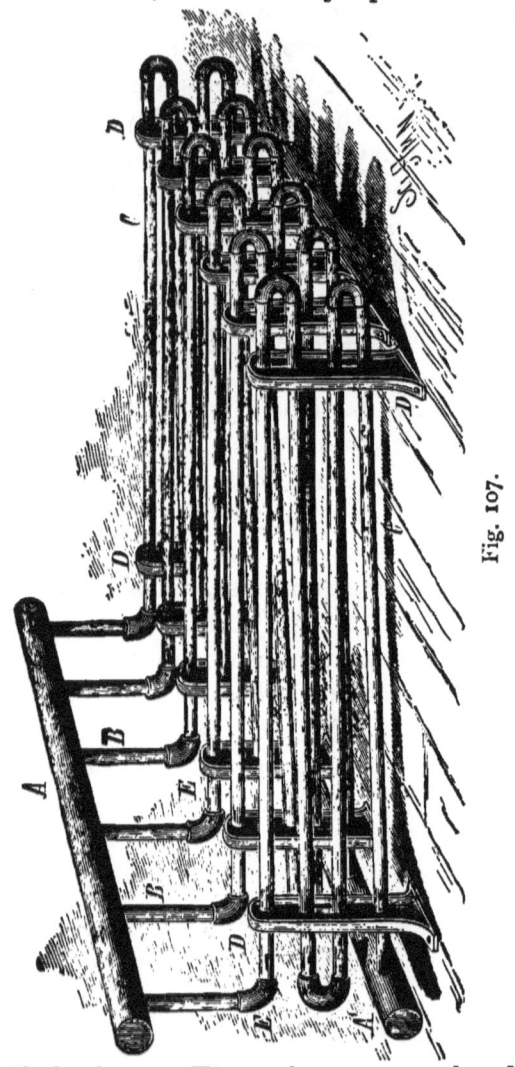

Fig. 107.

tion of all the heat. Thus, the greater the difference in temperature and the slower the movement of the air up to the point of saturation, the better the result in the laundry or dry kiln, or any place where rapid drying only is the object.

In no other place is the power of radiant heat (direct radiation) more manifest than in the drying room, and more failures can be traced to placing coils under skeleton floors, or flat on the floor, than any other cause, except, perhaps, an ignorance of the principles of piping, which so many consider can be done by any one who wears a pair of greasy overalls.

I have proved, in many cases, that the same amount of pipe or plate surface, distributed around and between the materials to be dried, will do the work in half the time it takes the heated air from an indirect coil. This is no mistake; and further, wooden blocks can be dried lighter (proving there is more water driven off) by direct radiation than by indirect radiation, the times and temperature being the same.

According to the above it is plain that in the construction of drying houses for most purposes, the heating surfaces should be so placed and distributed that the direct heat rays from the iron could fall uninterrupted on the greatest surface possible of the materials to be dried.

Fig. 107 shows a perspective of a good arrangement of a direct radiation laundry drying room coil, utilizing all the radiant heat that is thrown off and giving a thoroughly uniform heat throughout the room. A A' are headers (often called manifolds), usually made of extra heavy pipe to admit of tapping and threading instead of using T's, for the cost of the heavy pipe and the drilling and tapping is very much less, and the header better and straighter, than when composed of many short pieces of large pipe and the necessary T's. (These remarks apply to all large coils.)

B B are the spring pieces, threaded right and left handed; *C C*, the leaves or sections of the coil; and *D D*, the coil stands. The stands are always in pairs, to admit of giving the necessary division and inclination to the pipes, and when viewed through the holes look like Fig. 108. The dotted lines are the centers of imaginary pipes to show the pitch. When coils are very wide in the direction of the length of the headers it is well to keep the coil stand 2 or 3 feet from the header at that end, to prevent the expansion from

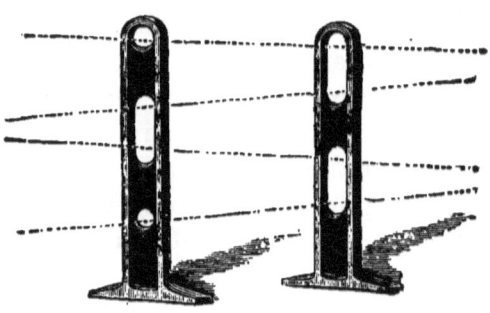

Fig. 108.

pulling the screws from the floor. The distance between the holes in the standing coil header is usually about 12 inches, or as wide as the clothes-horses are from center to center.

The usual way to build these coils is to start at the bottom header *A*, Fig. 107, and to put each leaf, *C*, together continuously, working upward until the elbow, *T*, is reached; when all the leaves are so far constructed, with all the elbows looking up, with their left-handed thread uppermost, count in and mark the right and left-handed spring-pieces B, then apply the upper header A, and screw the whole up as nearly alike as possible. Do not be persauded to do away with the spring-pieces and the elbows through economy, so as to connect the upper headers straight, as in a box coil; if you do you will have trouble should you want to take down a single leaf for repairs.

Fig. 109 shows sectional perspective view through a laundry drying-room; a being the coil; b, the clothes-horse; c, the suspended rail from which the horses hang, d, fresh air inlet duct; e, its damper or regulator; f, ventilator with regulator, usually governed by a cord and bell crank, and drawn back by a spring; and g, the space into which the horses are drawn, which of necessity must be as long as the horses. This style of drying-room gives the direct radiation from both sides of the leaf of the coil to the fabrics to be dried, and also exposes both sides of a fabric to the direct radiation of a section or leaf.

For high or low pressure steam 1-inch pipe is generally used in the coil; and if exhaust steam is to be used the pipes should be not smaller than 1 inch, and the total length of any one leaf should not exceed 100 feet of one-inch pipe under a back-pressure of 1 pound at the engine.

For exhaust steam the upper header should be large, 3 inches for 12 leaves of 60 to 80 feet each, or about 700 to 1,000 feet in the coil gives satisfactory results. This should be increased in porportion to the increase in the number of leaves, a 4-inch pipe header being enough for a coil of from 1,500 to 3,000 lineal feet of 1-inch pipe.

Fig. 109.

Unless the exhaust steam is carried a long distance horizontally, the pipe leading to the header may be

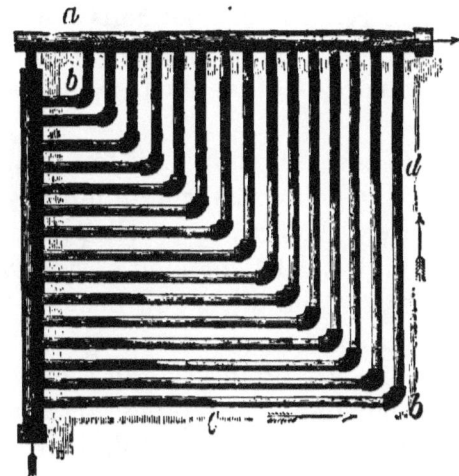

Fig. 110.

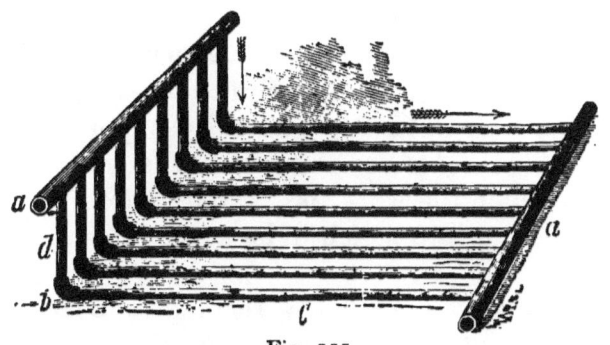

Fig. 111.

one or two sizes smaller than the header, provided it is large enough for the engine.

With steam of high tension, small pipe headers with

T-fittings may be used; but where the pressure is variable, a large header insures an equal distribution of steam to all the leaves of the coil.

Sometime gridiron or floor coils are used on account of saving expense, but the same amount of pipe in this form will not dry clothes as fast as the standing section coil.

Figs. 110 and 111 show gridiron coils of easy construction, $a\,a$ being the manifolds or headers; $b\,b$, right and left elbows; $c\,c$, coil pipes right handed, and $d\,d$, right and left handed spring-pieces. In Fig. 110 the pitch of the pipes and headers is in the direction of the arrows.

These coils are often used in lumber-drying kilns, but the same amount of pipes arranged around the walls in miter or wall coils will give a far better result, and will not be a receptacle for dirt, as a floor coil is, requiring a skeleton floor over it to walk on and pile the lumber on. There are places, however, for this kind of coil.

In piano-case manufactories, and where specialities in glued or veneered furniture of the best quality are made, the workmen are generally supplied with a drying cabinet, of a size suitable to the pieces to be dried, in which the work is heated before the glue is applied, and into which it is again placed to dry properly.

These cabinets are usually rectangler boxes, with holes in the bottom and top to allow the air from the room to circulate through them so as to carry off the moisture. Their steam coils are usually of the gridiron pattern, flat on the bottom of the box, with the valves on the outside. Sometimes they are heated indirectly by the warmed air conveyed in tin pipes from a large coil placed in some favorable position. Some manufactures claim the quicker the work can be dried after gluing the better it will be.

In large drying kilns on the direct radiation principle, where pipe enough cannot be put on the walls, rows of stancheons should be put up to hang the coils on, in such a manner as not to interfere with the gangways.

The tobacconist prefers to dry without artificial heat, in a temperature of about 60°, with a rapid change of air through the windows. This appears to give dryness without brittleness, but at night and in damp weather it is necessary to close the windows, and to get the stock out in time recourse must be had to steam coils. A temperature of 130° is generally considered ample, and can be easly attained in a drying room,

The additional quantity of pipe necessary to raise the temperature of a drying room from 120° to 130°, if again added, will not raise it from 130° to 140°. As the temperature of the drying room approaches the heat of the steam pipe, the heating surface has to be enormously increased to obtain an appreciable increase of heat in the room.

With low pressure steam—2 pounds per square inch or thereabouts—it is difficult to obtain a temperature above 175° in the drying room, no matter how much surface is used, and with steam at 60 pounds pressure the practicable limit is about 275° Fahr.

Before leaving the subject of drying-rooms I will add a little data I once obtained by experiment, when about to construct a large drying-room in a prison. It was neeessary to find how much water was driven off in the drying-room and to do so I weighed the clothing, dry before washing, and also weighed it after through an ordinary rubber wringer.

	Dry.	Wet.	Water to be driven off.
Socks (woollen)........	3½ oz.	9¾ oz.	6¼ oz.
Cotton shirt...........	12½ "	28 "	15¾ "
Undershirt............	12 "	26½ "	14½ "
Drawers.	10¼ "	21½ "	11¼ "
Two sheets............	25 "	50 "	25 "
Pillow-slip............	.....	12½ "	6½ "
Blanket...............	54½ oz.	129 "	74½ "
Towel................	4 "	8½ "	4½ "

CHAPTER XXV.

DRYING BY AIR CURRENTS.

When evaporation is produced by the direct action of steam, as in boiling water, it is a well established fact that to drive off 1 pound of water from any substance requires the heat of 1 pound of steam. This fact forms the basis of computation in arriving at the cost of drying. Of course there are other losses due to transmission, improper apparatus and wasteful traps, all of which must be added when know, nor assumed when unknown.

A maker of bricks once came to me for advice as to the best method of drying them in winter time, saying he had been trying some experiments with direct radiation steam coils placed under and around the stacks of compressed clay. He found the cost was enormous and probably prohibitory and wanted to know what, if anything, could be done to reduce the cost. He explained that his output was 60,000 bricks a day and that each brick contained an excess of about $1\frac{1}{4}$ pounds of water, or 75,000 pounds of water he had to get rid of each day.

His boiler was about 100 horse-power, as boilers are roughly rated in boiler makers measurement, and he had run it day and night, burning 5 to 6 tons of coal at $5 per ton, without making any very appreciable impression on the moisture in the bricks over what they could show without any heat. Now this person was evaporating about 5,000 pounds of water every hour in his boiler, and for 24 hours it would amount to 120,000 pounds, or about double the water he was trying to drive from his bricks. Still he was not successful because the heat could not be applied in such sheds, and even with the best of sheds he probably would have to evaporate twice the quantity of water in the boiler that he could drive from the bricks by the condensation of the steam. What was to be done? The answer was, try air. Warm dry air if you can get it, if not, such air as you can obtain.

If air is warmed from 50° Fahr. to 100° it is capable of taking up approximately 15 grains of moisture per cubic foot of air. Of course the air will be cooler somewhat by this, but assuming it is still capable of taking away only 5 grains of moisture per cubic foot, taking into consideration it is cooled to 50° Fahr. in doing so, 1,000,000 cubic feet of it is capable of taking away 5,000,000 grains or 1,000 pounds of water. Now it was necessary to take away 75,000 pounds of water in 24 hours so that approximately 3,000,000 cubic feet of air per hour under these conditions properly applied will dry the bricks.

The expense will be the cost of moving the air with a fan and engine, which under a resistance equal to $\frac{1}{4}$ inch of water per square inch of surface, is theoretically equal to 1 horse-power per million cubic feet of air moved, and in practice is about 6 horse-power for 3 million cubic feet. The exhaust steam from this engine and from any other source or engines there may be on the premises can be used to warm the air as far as it may go, but I do not propose to warm the air by direct steam from the boiler, as almost any air, day or night, is capable of taking up some moisture, as every one should know by watching clothes dry on the line during a freezing day, or any day that is not raining or foggy. Five grains of moisture per cubic foot of air is probably not too high, but even if it is, by 50 per cent. twice the quantity of air moved will do the same work, so that 12 horse-power for moving air is a liberal estimate for drying 60,000 bricks. It is simply accelerating Nature's method.

Since that time many persons dry bricks with force and a little heat, particularly stove linings, terra cotta and the like. The goods are placed on racks in long rows enclosed by shutters, so the air will be forced lengthwise over the whole mass. Sometimes, where engines are not wanted, aspirating chimneys are built and the air is moved in this way by drawing it. Fig. 112 shows a method where the fan is employed, and Fig. 113 shows the aspirating shaft.

When steam is used coils may be placed, as shown, in the racks. When the air first enters it has a certain

Fig. 112.

272   STEAM HEATING FOR BUILDINGS.

capacity for moisture, but of course this becomes less as it travels on its way over the wet goods. A coil at *a* raises the temperature of the air, and its power

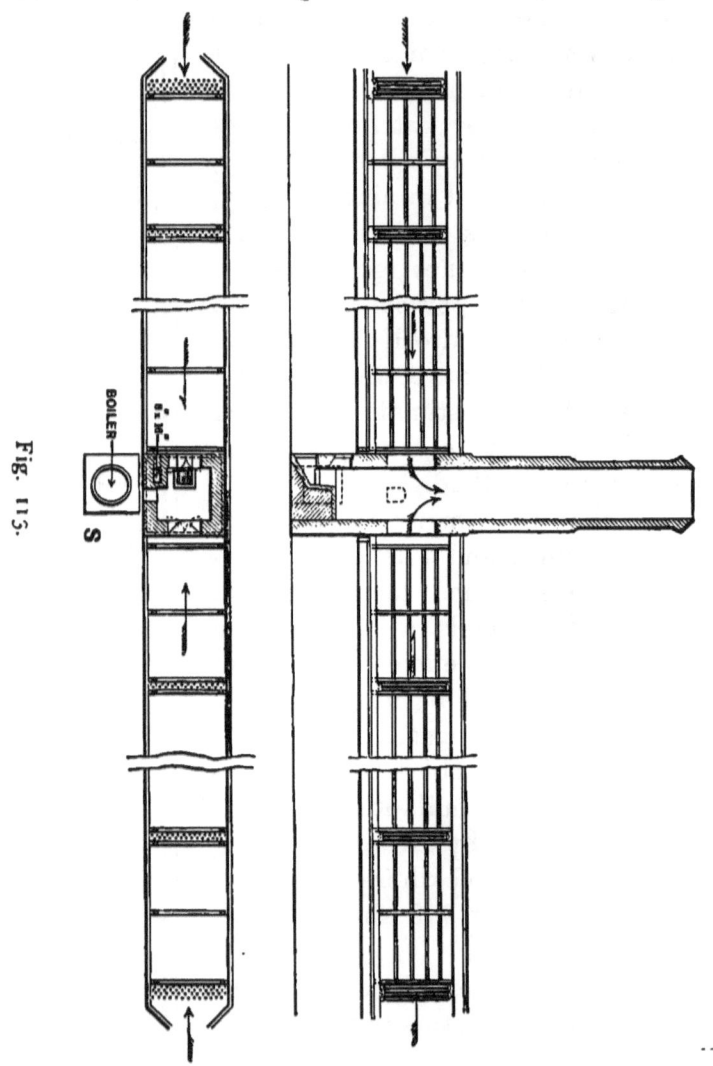

Fig. 115.

to take up more moisture is increased. As it becomes saturated again, another coil, *b*, helps its absorbing power again, and this may be carried to a considerable extent in drying certain things. It produces a comparatively uniform drying and prevents the necessity of changing the positions of the goods. In the case of brick or heavy articles that cannot well be moved, ducts may be run underground or in any other manner whereby the air can be diverted so as to blow for an hour from one direction and an other hour from another direction, or from different directions, modifications of which will suggest itself to any inventive mind.

Innumerable are the uses to which modifications of this system can be applied. I have used it for drying bricks and stove linings and also to dry shoes in certain stages of their manufacture.

In Fig. 113 *s* is the boiler to make heat in the asperating shaft or chimney.

When it is not desirable, or it is impracticable, to use relay coils in such drying rooms, it often happens the goods can be stacked on movable racks on wheels so that by changing positions evenness in drying can be obtained.

It is well to call attention also to the fact that goods dried in a high temperature will appear damp, and will be actually damp when removed and allowed to cool to the temperature of the outer room. This is caused by the great quantity of moisture, warm or hot air may contain and still be relatively dry. Airing before a hot coil or stove will drive off this residual moisture.

CHAPTER XXVI.

STEAM-TRAPS.

A STEAM-TRAP is an appliance attached to certain classes of steam apparatus, whose object is to remove the water of condensation without a waste of steam. A gravity apparatus does not require a steam-trap of any kind; and a proof of a perfect gravity circulation is shown by the proper working of the apparatus without one.

Traps may be separated into two principal classes —namely, traps which open to the atmosphere or into a receptacle having less pressure than the heating apparatus, and direct return traps, returning the water to the boiler, without a considerable loss of heat or any loss of water.

I will speak of the gravity return trap first, though they are now almost superseded by pump governing apparatus where there is sufficient pressure in the boiler to run a pump.

These direct return or automatic traps came into use about 1870, and then formed a new departure in steam-traps. They must, to be efficient, be automatic in action and of simple construction and positive, for an interruption of an hour will fill the coils and pipes with water, and in very cold weather may be the cause of freezing the apparatus, so that caution must be exercised in the selection of them. There are now two or three very good modifications of this trap before the public. They are not suitable, however, under the practice of the present time, except as an auxiliary to a gravity apparatus where the coils are below the water line and where the pressure is too low to run a pump controlled by an automatic governor regulated by the flow of the return water.

The principle involved in these traps is simple, being alternately a vacuum and a pressure, but, like the single acting reciprocating pump which has no flywheel to help it at the end of the stroke, it must have some kind of an auxiliary.

With the aid of the diagram, Fig 114, the action of these traps may be explained. A represents the boiler; B, the trap proper; C, the receiver, which holds a certain quantity of the return-water; D, a steam pipe from the boiler to the trap; E, a pipe from the trap to below the water-line in the boiler; and F, a pipe from the receiver to the trap carried up inside the globe. It will also be seen these pipes are provided with valves; the steam-pipe has a globe-valve, and the other two pipes, check-valves; the valve in the pipe F, opening toward the trap, and the valve in the pipe E, opening toward the boiler.

Now, if the valve in the steam-pipe is opened and steam admitted to the globe B, until all the air is expelled and the steam allowed to condense, as it will do in a short time after the valve is closed (by the loss of heat from the steam through the sides of the globe to the outside atmosphere), there will be a vacuum formed in the globe, more or less perfect, which will draw water from the receiver *C*, when there is a pressure in the pipe which comes from the coils or elsewhere; and this water, passing the check-valve in '*F* will overflow into *B* and cannot return to *C*, for two easons—because it cannot pass the check-valve back-

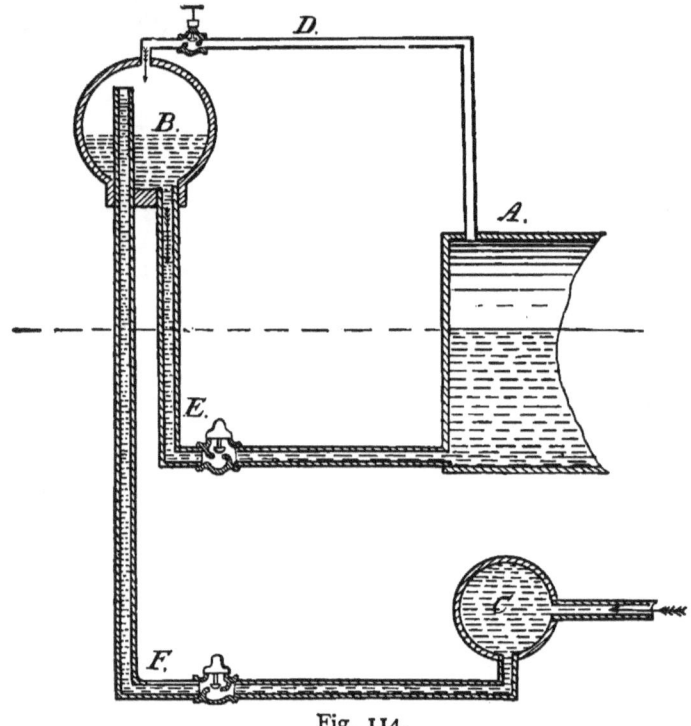

Fig. 114.

ward and cannot get back over the top of the pipe *F*. Now, if the valve in the steam-pipe is opened and the pressure of the steam in the boiler admitted into the top of the globe *B*, the pressure will become equalized between the boiler and the globe, and allow the water to pass down the pipe *E*, and into the boiler of its own gravity (precisely as it would if everything was opened to the atmosphere), going through the other check-valve, which will not allow it to pass back again when the valve in the steam-pipe is closed. Condensation will again form a vacuum, which will once more draw the water from the receiver to flow down into the boiler when the steam-valve is again opened, and thus the action goes on, being simply that of a pump without a piston. I say the vacuum will draw the water from the lower bulb; this of course is not exactly what goes on. It is the pressure that sends it up, and as soon as the trap grows a little old and its steam valve leaks, as it always does, it would not work at all were it not for the pressure in the return pipe.

This principle was understood and used substantially as explained above before the automatic traps were introduced, but as it was necessary to construct the two globes or tanks of large size to avoid too frequent attendance, and as it required manipulation at irregular intervals, which, if neglected, would fill the pipes with water, it was not much used. Now, since automatic contrivances have been invented, which take the place of manipulation, and which can be depended on with some degree of certainty, these traps can be, and are, used on apparatus which otherwise would be almost useless. Thus the difficulty to be overcome in this class of traps as before mentioned is to construct an automatic contrivance for opening and closing the steam-valve, which can be relied on.

Fig. 115 shows one of these traps, which has been selected as an example, not because the trap is considered the best—for there are others equally good—

but because the action of the auxiliary is so easily explained. It is a view of the trap when set up; *H* is the steam-pipe; *G*, the pipe from the receiver to the trap; and *F*, the pipe from the trap to the boiler. The valve marked *D* is the steam-valve, which is automatically regulated, and is a rotary slide-valve; *E*, a connecting rod between a crank on the valve stem and an

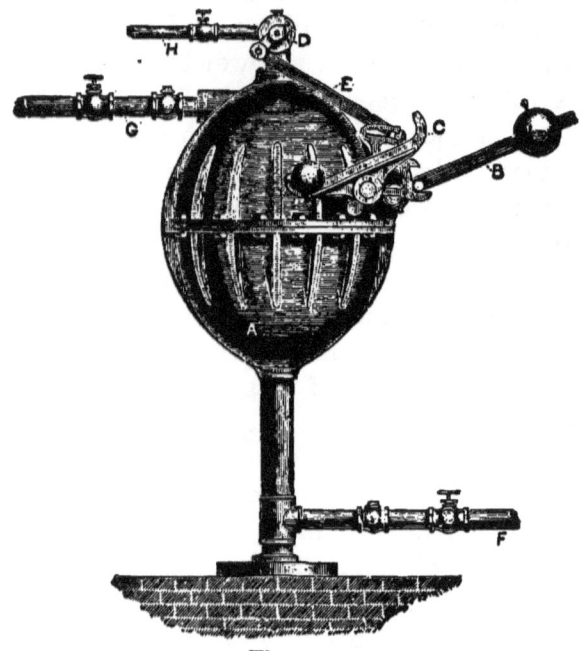

Fig. 115.

arm with slack motion, a part of the casting *C*, which rocks on the stud; *C*, a track, on which rolls a ball, also a part of the casting which rocks on the stud before mentioned, and which engages another stud, on the lever *B*; the lever *B* and its weight are a counterpoise to a float inside the globe. The action is as follows: When there has been vacuum in the globe, the water will pass through the pipe G, and fill the trap, consequently it raises the float and lowers the lever and counterpoise, whose stud engaging C, draws it down until the track passes the horizontal position, without affecting the connecting rod E, on account of the lost motion. When the track has passed the horizontal position, the ball will roll along the track and strike on the opposite end against the hook, giving a blow sufficient to move the valve on its seat and open it to its full extent, but not before the globe is full of water. The reverse motion is similar: the float lowering, but not affecting the valve, until the water is nearly all out of the globe; the slack motion allowing the valve to remain open until the track again passes the horizontal position, when the force exerted by the blow on the hook at the other end of the track closes the valve suddenly.

Among the atmospheric traps are found the old expansion traps, now little used, and the open float traps, which still form a necessary part of certain apparatus.

Cooking apparatus, such as meat-kettles, or kettles or tanks with coils in them, which condense much steam in a short time, should not be connected with a low pressure gravity apparatus but should have a separate pipe from the boiler, and be connected to a trap, in consequence of the great and sudden shrinkage of steam, which takes place when they are quickly filleb with cold water.

Fig. 116 shows a well known type of open float-traps, used both in this country and in England, of which there are many modifications of minor importance;

the action and principle remaining the same. *A* is a cast-iron pot, sufficiently strong to withstand high-pressure steam, with an inlet at *F*; *B* is another pot (an open pot), inside the pot *A*, with a spike at the center of the bottom and a guide to keep the inner pot in a central position. *C* is a brass tube screwed into the cover *D*, and forming a valve with the spike at the inside of the bottom of the pot *B*; *E* is a valve in the cover of pot *A*, which, when opened, acts as

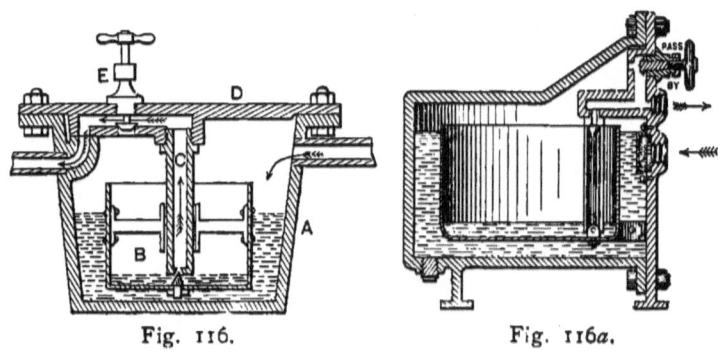

Fig. 116.   Fig. 116*a*.

an air-valve, or *blow-through*, to quicken the circulation when first turning steam on the spparatus.

The pot trap operates thus: the condensed water from the coils, etc., runs in at the pipe *F* and fills the outer pot *A* with water until it floats in the inner pot *B*, against the stem *C*, closing the valve formed by the spike and the tube, thus closing the outlet to the tank or the sewer. The water, which still continues to flow into the outside pot, rises and overflows into the inside pot. Then the latter sinks and opens the valve which the spike forms with the hollow stem and allows all the water in the inner pot to be forced up through the stem and out by the pressure of the steam in the upper part of the pot acting on the surface of the water. Thus, when the inner pot becomes buoyant again, by the discharge of its water, it closes the valve and leaves it so, until the increase of the condensed water again overflows it. This action is intermittent, the frequency of it depending on the amount of work to be done.

There is one point in the construction of this trap on which its working depends, namely, the area in square inches of the hole in the end of the hollow stem C must be no larger than the quotient obtained from dividing the weight in pounds of the inner pot when submerged, by the maximum pressure in pounds per square inch of the steam to be carried. Thus, if the inner pot weighs $12\frac{1}{2}$ pounds under water, and the greater pressure of the steam is to be 100 pounds per square inch, the whole must be a little smaller than $\frac{1}{8}$ the area of a square inch, say a round hole $\frac{1}{4}$ of an inch in diameter, which leaves a factor of safety of $\frac{1}{8}$ the weight of the pot. The reason for this is plain when we consider that there is practically no pressure within the stem when the valve is closed, and for the pot to sink, when it is full it must be heavy enough to pull itself away from the stem, and still be light enough to float with considerable buoyancy when empty.

This *type* of trap possesses a special point of excellence; it will discharge the water of condensation from coils or from the cylinder of an engine into a tank or sewer at a very much higher level than that which drains, and it will keep them as dry as if it discharged downward. It is peculiarly adapted to elevator engines and pumps which stop and start frequently, and are operated from the car or an upper floor, as it removes the water at a high temperature and will keep a steam chest and cylinders dry by removing the water which accumulates while the engine or pump is standing with steam turned on.

The best known modification of this trap is the "Mason." Mr. Charles E. Emery, C. E., designed a modification of this trap which works with a slide valve, the object undoubtedly being to get a trap that would open under high pressure and give a compara-

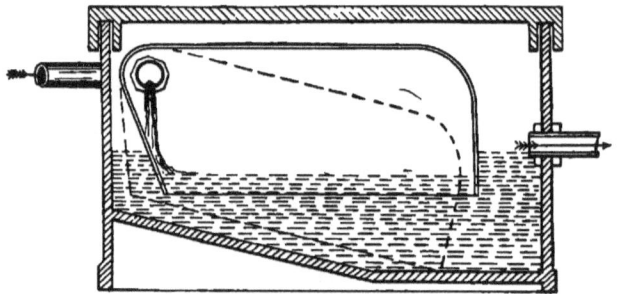

Fig. 117.

tively large opening. Another modification of the trap in which leverage is employed for the same purpose is the "Kieley," Fig. 116a. There is another open float trap, Fig. 117, which contains a special point of merit, the valve of which is not much understood, namely, a trap capable of taking recognition, so to speak, of temperature, as well as quantity, and which will discharge its water down to atmospheric temperature and pressure, no matter what may be the temperature of the water in the coils due to high pressure.

To make this clear, it is necessary to explain that

pressure, leaving the temperature of the water (when water which falls to the bottom side of a nearly horizontal pipe with 50 pounds pressure of steam in it, has not fallen to a temperature of 212° Fahrenheit, as is very generally supposed, but has simply parted with the latent heat of the steam, incidental to the the flow and pressure of the steam are maintained) a very little less than the temperature of the steam. When this water is exposed to a lower pressure some of it will evaporate again, and part of the sensible heat of the water will become latent in the steam thus formed. But it must not be understood that all the water flies into steam. It does not; the quantity of water converted into steam being represented by the ratio, the latent heat of the steam at the different pressures, bears to the sum of the latent heat and the sensible heat of steam. Thus, when water is drawn directly from a high-pressure coil into the receiver of a trap, and is discharged against the presence of atmosphere, before the water has cooled below 212°, that is a considerable loss of heat. This can be seen in the blowing of a gauge-cock, for though the water is solid and dense in the boiler, when it is drawn, some of it flies into steam and makes a cloud which often deludes the novice into the belief that it is all steam, and that possibly he has low water. The construction of this trap is plain; it consists of an outside case with a loose cover, an open float with the mouth down, and a common plug-cock operated by the float. When steam or water above 212° in temperature is discharged through the cock and under the float, the latter is immediately raised by the pressure of the vapor underneath and between the float and the water which cannot flow over the case. This action closes the cock, which will remain closed until the vapor condenses and allows the float to once more sink, when the cock again discharges the hot water behind it. If this water is below, 212°, it will pass rapidly out of the case under the edge of the float; but when it again becomes hot enough to make a little steam, the float raises and the cock is again shut.

This trap cannot be used on an engine, as it will not discharge any considerable quantity of water until the temperature is below 212°; but for an expansion system, where the trap has not to discharge against pressure, or for an exhaust steam system, it is a good one.

There are many forms of expansion traps or those depending on the difference of the expansion of metals for their operation. They are difficult of adjustment, and are apt to be erratic in action, as they change their length with every change in steam pressure.

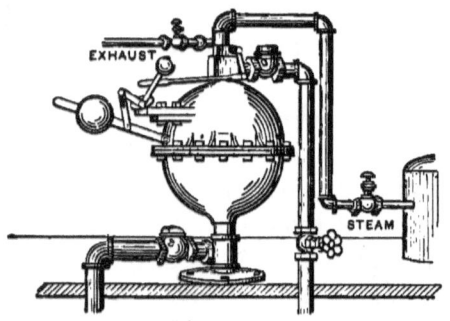

Fig. 115a.

Fig. 115a shows the best known form of the type of trap shown in Fig. 115. It is the Kieley "Champion" and depends for its positive action on a weighted lever that alternately passes its centre of gravity.

CHAPTER XXVII.

VALVES FOR RADIATORS.

155. It is not necessary that I should say much about radiator-valves, as all practical men must be conversant with their details. In an early chapter of this book, I pointed out the objection to using a globe-valve except on its side for any purpose, and in connecting radiators I am not sure but that I should take the position that a globe-valve should not be used under any circumstances. A suitable straight valve for a radiator we have not, as a gate valve is unsuited for high pressures. The angle valve, therefore, becomes the favorite valve for radiators, and in most cases it can be used with the pipes coming through the floors, and the steam and handle of the valves uppermost. It is, however, of irregular cases that I have to speak, and I find that it is often better in matters of this kind to warn a person against what should not be used, than to say a great deal about what may be used. Therefore, I say, never use a globe valve as a radiator valve when you can avoid it, and never under any circumstances use a globe valve as a radiator valve with the stem and handle in the vertical position.

286 STEAM HEATING FOR BUILDINGS.

Fig. 118 shows a radiator connected with two globe valves and it is only necessary to have a practical man study the drawing and discover the mistake that is so often made without my referring more to it in detail.

To those who have had little or no experience with heating I will explain that there is practically as much pressure in the return end of a radiator, in an ordinary system, as there is in the steam end, the two globe valves (as shown) will "trap" the base of the heater full of water, as represented. When condensa-

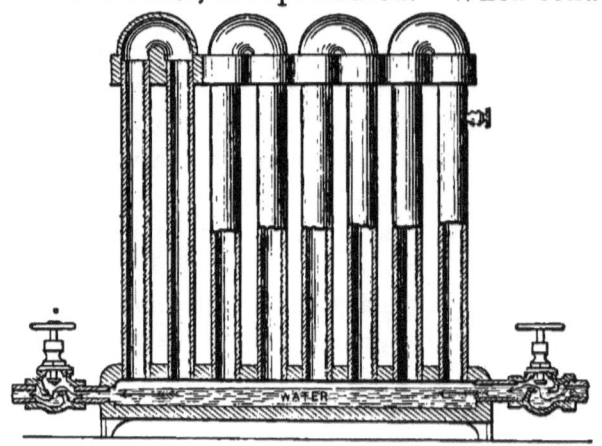

Fig. 118.

tion takes place, as it must, within the pipes of the radiator, the greater pressure in the pipes will force inward at the two ends and nearly alike. This will prevent the water accumulated in the base above the valve line from flowing out easily, as it should, and will make it assume a level still higher than is shown. This will make the water rise against the bottom of the tubes, and allow it to pass up in the tube the air valve is on when the latter is opened by relieving the pressure in the heater still more. The natural result is the partial filling of the radiator with water, the accumulating column of which preponderates against the entering steam at the lowest or return end intermittingly, thus finding its way out of the heater sufficiently not to let the heater become cold, but accompanied by noise.

If the valves are turned the conditions are somewhat better, as then the steam has a clear passage, though if the globe valve must be used it is better to turn them with the stem sidewise, but not quite down to the level. This gives a clear and level waterway on a vertical section

When the radiator connections are above the floor, the angle valve cannot be used unless the radiators have unusually long legs, or are set on supports, both of which are objectionable. The *ordinary* corner valve, which is no more than a globe valve with the inlet and outlet of the valves at right angles to each other, are just as objectionable as the globe valve proper, and in fact more so, as they cannot be used on their side as the globe valves can, so that this brings us to the use of the offset corner valve shown in Figure 119.

Offset opposite valves are not new, although it is only recently that any maker has kept them in stock.

Lewis Leeds, the ventilating engineer, used them fully twenty-five years ago, and H. M. Smith, an engineer of the New York Steam Company, now deceased, modeled the offset corner valve after Mr. Leeds' valve so as to make a neater and shorter connection without the use of a nipple and elbow. His object was to provide a corner valve for use when connections to risers must be made above the floors, that will admit of the running of all the water of condensation from the base of a radiator by gravity, and which in the ordinary globe valve is forced to rise to the height of the seat or "bridge" when the stem is in an upright position, as shown in the illustration of the radiator Fig. 118.

To accomplish this an offset is formed in the body of the globe, which is plainly shown in Fig. 120, and qual to a little more than one-half of the diameter of the pipe. The inlet (when the valve is used as a "re-

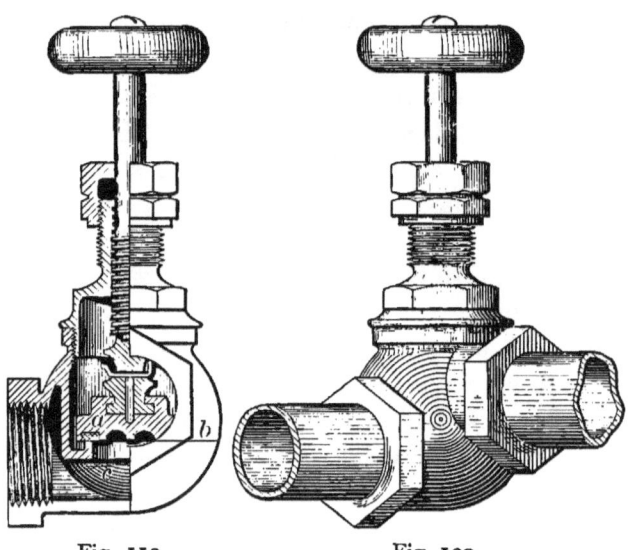

Fig. 119.   Fig. 120.

turn-valve") enters highest, and its lowest side is level with the top of the seat, over which the water must flow to pass into the outlet. When it is used at the steam end of a radiator, the lower pipe then becomes the inlet, and water formed in the nipple can run backward to the riser if it is not carried into the base of the radiator.

Fig. 119, partly in section, shows this arrangement, *a* being the valve-disk, *c* the valve-seat, and the line *b* the level of the top of the seat.

The offset corner valves are made "right hand" and "left hand," so as to be used on opposite ends of the same radiator, and in general practice the offset valve whether opposite or otherwise, can be used in view of globe valves without materially increasing the length of the legs of the radiators when they are short, and with most ordinary radiators as they are now built, with the inlets and outlets about four inches from the center to floor.

Fig. 121 shows the second of these improvements. It is really not a modification of a radiator valve, but an improvement that may be added to any valve that is to be controlled from a distance. Its application to a radiator valve is directly for the purpose of controlling the steam in the radiator by the temperature of the room, the medium being a metallic thermostat operating an electric or pneumatic circuit, which in turn operates the valve by pneumatic pressure. It is called the Johnson Pneumatic System. Compressed air is used as a motive force for opening or closing valves which regulate the heat supply. The compressed air is obtained either by an hydraulic or steam compressor, and is stored in an air receiver, and thence distributed by suitable pipes to the thermostats, from which in turn the air is carried to the valves.

The steam or hot-water radiator valve has an ordinary valve body set in the usual manner at the end of the radiator, and connected with an expansive diaphragm which serves to open or close the valve. This will be better understood by the following sectional view of the diaphragm valve.

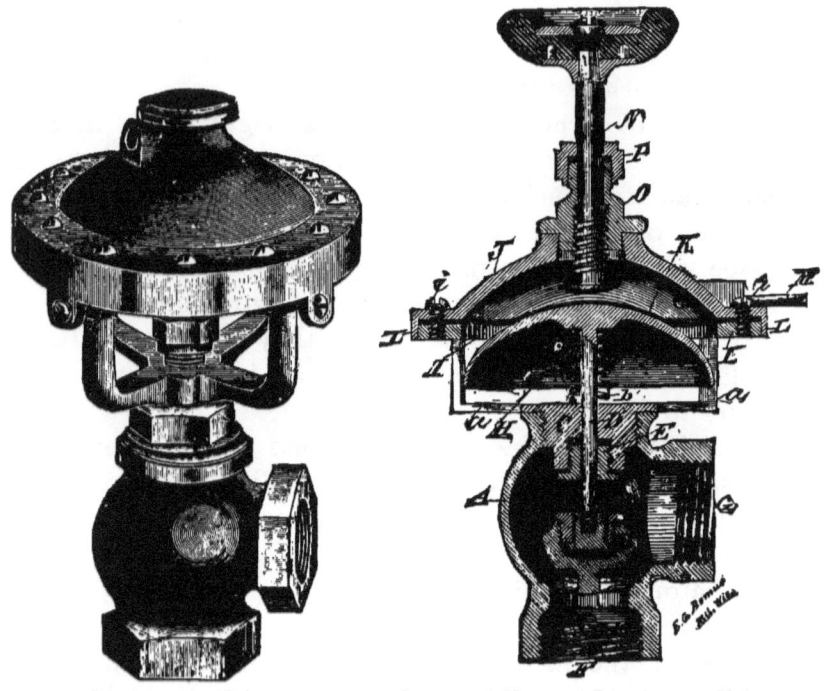

Diaphragm Valve.   Sectional View of Diaphragm Valve.
Fig. 121.

A is the valve body; B, the valve disc; C, the packing box through which the stem passes; H is a saucer-shaped piece fastened to the upper end of the stem D. The valve is held open by the steel spring b, which presses upward on the saucer H. Above this saucer H is the umbrella-shaped piece J, held by the standards a, a. Upon the under side of the piece J, and fastened to its edges to produce an air-tight joint, is the flexible diaphragm K, made of cloth and rubber. There is an opening through the pipe M into the chamber formed between the metal piece J and the diaphragm K. When air, under pressure, is admitted through the opening M, the valve will be pushed downward to its seat. When the air is allowed to escape from above K, the spring b will open the value B to its full extent.

The passage of the compressed air to the diaphragms on the valves or dampers is controlled by thermostats located in the rooms where the temperature is to be controlled.

The thermostat consists of a compound strip, made of brass and steel, and a small double valve. It is provided with an index whereby it is set to operate at any reasonable temperature, and can be so accurately adjusted that it will operate on a variation in temperature of one degree. A thermometer used on the face is merely for the purpose of showing the temperature and testing the accuracy of the regulation.

CHAPTER XXVIII.

REMARKS ON BOILER CONNECTIONS AND ATTACHMENTS.

Feed Pipes.—The feed-valve should be a globe or angle valve placed near the boiler, with the fewest possible joints in the feed-pipe between it and the boiler. If it is a loose or swivel disk valve, it should be secured with solder (sweated in) in the threads of the double part of the disk, so as to make it almost impossible to loose the disk from the stem; a mark with a center punch or chisel is not enough. The valve should be so turned toward the boiler that the inflowing water will be under and against the disk, so that in the case of the loss of the disk, it will not act as a check-valve against the influx of the feed-water. This arrangement will bring the pressure of the water in the boiler always against the stuffing-box of the valve; but all things considered it is best.

The check-valve should be closed to and outside the feed-valve, with only a nipple between them. Always use horizontal check-valves, as they admit of easy cleaning. With the ordinary vertical check it is necessary to take down some part of the feed-pipe to clean it.

When two or more boilers are fed from the same pump, or when the pump is used for pumping water for some other purpose, it is well to have a stop-valve on each side of the check-valve, as it will enable the engineer to get at his check without stopping the water elsewhere.

In passing through boiler walls or cast-iron fronts, care should be taken that the feed-pipe does not nest, or the settling of the boiler will break it off.

Use a flange union on the feed-pipe instead of the common swivel union. The engineer can take a flange union apart with a monkey wrench, and it makes a more permanent piece of work than an ordinary union and is not likely to leak.

Never put a T in the feed-pipe inside the feed-valve for the purpose of a blow-off; but make a separate blow-off connection to the boiler.

Blow-off Cocks.—Never use anything but a plug cock of the best steam metal throughout or the asbestos cock. The reasons for using a cock are, that the engineer is always sure when he looks at it whether it is shut or open. It gives a straight opening. If chips, packing, or dirt gets into the cock it will shear them off when closing, or if it does not, the engineer knows it is not shut. Do not use an iron-body cock with brass plug, for when the cock is opened to blow down a little, the hot water expands the plug of the cock more than the body, and it is almost impossible to close it. Do not use a globe or angle-valve, as you cannot always tell when it is shut; a chip or dirt getting between the disk and seat will prevent its closing. I have seen two fine boilers destroyed from this cause. Gate or straight-way valves are subject to the same objections as globe or angle types.

When it is practicable there should be a T with a plug in it in the blow-off pipe outside the blow-off cock, and the plug so arranged as to be removed when the cock is closed. By this means the engineer can always tell if he is losing water from his boiler through the blow-off, and in the matter of expert trials it will not be necessary to remove a section of the blow-off pipe, as the engineer can satisfy himself through this hole as to whether there is a loss from the boiler or not.

The blow-off pipe should be large, with few bends in it, and fire bends are better than elbows. It should be attached to the bottom of the shell of a horizontal boiler, and not tapped into the head a few inches up. When there is a mud-pipe, attach it at the opposite end from the feed-pipe.

Safety-Valves.—They are the main-stay of the engineer, acting both as a relief and a warning signal. They should be attached to the steam dome high up. The side is better than the top, as they are not so liable to draw water when blowing off in that position. They should be large, and have a large pipe connection all to themselves if possible. The ordinary cross-body safety valve, when used as such, is very much to be condemned, and I think in some countries there are regulations against their use. They are constructed to save making an extra connection for the main steam pipe, thereby drawing the largest amount of steam directly from under the disk of the safety-valve. A weighted safety-valve is better than a spring-valve when it can be used, as the lifting of the valve makes practically no difference in the leverage ; not so with a spring-valve, for the higher it is lifted the more power it takes to compress the spring.

Gauge or Try Cocks.—Gauge-cocks are various in style, the wood handle compression gauge-cock is a very good kind for all purposes. When setting gauge cocks care should be taken that they are not too low, and that the drip will not flow over the person who tries them. They should be tapped directly into the boiler if possible ; but when it is necessary to use a piece of pipe to bring them through a boiler front or brick work, give the pipe an inclination backward, that the condensation may run back and into the boiler. When the pipe inclines outward and down, the condensation remains in it and the cock, and will deceive the unwary, giving the appearance of plenty of water with a short blow.

Glass Water-Gauges.—Water-gauges are best set when attached to a vertical cylinder at the front of the boiler. The cylinder should be connected to the boiler with not less than 1-inch pipe, top and bottom ; the top or steam connection should be taken from the boiler shell near the front head, and not from the dome or steam-pipe, as the draught of steam in either will cause the glass to show more water than the boiler contains. The bottom or water connection should be taken from the front head at a point where about two-thirds of the water in the boiler will be above it and one-third below ; this will lessen the chances of the pipe stopping up with mud, etc., and it should also be provided with a half inch pipe at the lowest point for a blow out. When gauge glasses are set this way the condensation in the cylinder is downward, and the flow of water being toward the boiler through the bottom pipe, the tendency is to clean the glass and cylinder and keep them so.

Steam Gauges should never be set much above or below the boilers to which they are attached, as each 27 inches of fall or elevation from the direct connection is nearly equal to a difference of one pound on the steam-gauge. It is always so when the gauge is below the point of attachment, for the condensation in the gauge-pipe fills it with water, which leaves a pressure on the steam-gauge equal to the hydrostatic head, which is a little over two feet perpendicularly to the pound per steam-gauge, giving the gauge the appearance of being weak. When the gauge is above it is not so always, though generally so even then, for the pipes being long and of small diameter, or trapped, which prevents a circulation of steam in them, they fill with water, which acts against the pressure from the boiler and gives a gauge the appearance of being strong.

When it is necessary to have a gauge very much lower than a boiler, fill the pipe with water, but before doing so remove the glass and lift the hand or index over the stop-pin and mark where it remains stationary. Now fill the pipe to its highest point with water, then draw the index hand from its spindle and set it back to the mark where it remained stationary before the pipe was filled, and press it on; then bring it to its normal position on the stop-pin and adjust the glass.

The Main Steam-Pipe for Heating Apparatus should be high up on a boiler, and any pipe larger than 2 inch should not be tapped in, but connected with a flange bolted or riveted to the boiler. Two-and-a-half inch pipe and larger sizes have eight threads to the inch, which forms too coarse a pitch to be tapped into one thickness of boiler-iron.

Automatic water-feeders, combination water-gauges, or steam-gauges, should not be tapped into the steam-heating or engine pipe, as the draught of the steam through the pipe interferes with their proper working.

Engine or pump pipes should not be taken from the steam heating pipe of a gravity apparatus, as the draught they cause relieves the pressure in the heating apparatus to a considerable extent.

All pipes connecting with boilers should be *extra thick* until at least the first cock or valve is reached, as much of the pipe now on the market is below the old Morris Tasker Co. standard, and is too thin, 1 inch to 2 inch inclusive being the sizes which give the most trouble.

CHAPTER XXIX.

DATA ON CONDENSATION IN RADIATORS.

Mr. Thomas Tredgold, early in the present century, considered the question of loss of heating and radiating surfaces in a very thorough manner, and later he was followed by Mr. Charles Hood on the same subject, both having the same object in view—namely, to find the value of radiating surfaces for warming buildings.

Mr. Tredgold found that 2.19 pounds of water cooled from 180° to 150° Fahr. in a vertical tin cylinder in 46 minutes, the exposed sides of which were 79 square inches, when the temperature of the room was maintained at $55\tfrac{1}{2}$° Fahr. during the trial. This gave a mean difference between the air of the room and the surface of the cylinder of 109.5° Fahr.

From this we have 2.19 pounds of water cooled 30° Fahr. by $\tfrac{79}{144}$ of a square foot of surface in 46 minutes of time, which is equivalent to 65.7 heat-units for the time, or 85.7 heat-units for an hour of time, and 156.21 heat-units as the quantity what would be given off by one entire square foot of the same surface (tin cylinder) in an hour of time. This total heat, for a square foot of surface, for an hour of time, then, divided by the mean difference of temperature (109.5 Fahr.) between the air and the surface of the cylinder equals 1.42 heat-units; the amount given off per square foot of surface per degree difference of temperature.

His second experiment was with a glass cylinder that held 2.125 pounds of water and had a surface of 71 square inches. It cooled from 180° to 150° Fahr. in 31½ minutes in a temperature of 56½° Fahr., which, by the same method of reasoning as we used before, gives 2.248 heat-units per hour per square foot of surface per degree (Fahr.) difference of temperature.

His third experiment was with a sheet-iron cylinder the surface being that of new sheet-iron unpainted whose surface was 76.7 square inches, holding 2.14 pounds of water and cooled from 180° to 150° Fahr. in 29 minutes, the temperature of the air of the room being 57° Fahr. By the same reasoning and method of calculation used in the foregoing examples we can find that the sheet-iron gave off 2.35 heat-units per hour per square foot of surface per degree difference of temperature.

These cylinders were as nearly alike as they could be obtained in form and size, and one cover fitted all. They were suspended by cotton threads, so that little or no heat could be lost by conduction or contact, and the sides and bottoms were exposed to the action of the air, etc. The top was covered by about one inch in thickness of alternative folds of cotton and flannel, so that the loss of heat by this direction was very small.

A few days later, when the experiments were repeated, the iron cylinder had become rusted. This

Mr. Tredgold says, increased its efficiency in the proportion of 156 and 180; the rusted cylinder having having the latter value, when as a new one it had the former. The experiments with the tin are of no value to us, except to show that bright surfaces have a less value than dull or slightly roughened ones. Experiments with brass, etc., by other experimenters confirm this. The relative value of glass and iron, however, are of some value to us as showing how nearly they agree; the iron being the better of the two, even when new and bright, and increasing in value as it becomes rusty.

It would be well to remark here that, probably, when surfaces become rusty, which they will in practical heating, they may deteriorate somewhat, and that it would be well to assume what they may increase in efficiency by rusting will be fully offset by accumulations of dust, etc.

The form of Mr. Tredgold's cylinders (short, vertical ones) are presumbly the best that can be devised for giving off heat. The same cylinders in a horizontal position would probably be found to be a little less efficient, and if they were to be increased in height, say two or three times, though used in a vertical position, it is only reasonable to suppose they would do *less* duty, for the very simple reason that the air in contact with the upper parts would have been warmed somewhat by the lower part as it passes upward, and, therefore, is not capable of extracting as much heat. The same holds good of horizontal pipes or cylinders when placed one above the other; each successive one, counting from the bottom upward, does less work than the one next below it.

According to the above relative values, therefore, of glass and iron, the empirical rule given before for finding heating surfaces by the window area, etc., is not without some scientific pretence, as the loss of heat through the glass of a window can rarely, if at all, be greater than through the iron of the heaters for equal difference in temperature, or for proportional differences.*

To go further with this subject, I will refer to experiments of Mr. Hood made more recently than Mr. Tredgold's, as he was not satisfied with the latter's deductions, and made experiments for himself. In his work on "Warming and Ventilation" he tells us that "to asertain the velocity of cooling for a surface of cast-iron, a pipe 30 inches long and $2\frac{1}{2}$ inches internal diameter, and three inches diameter externally, was used. The ends of the pipe were closed by corks, which entered the pipe $1\frac{1}{2}$ inches at each end, and the bulb of the thermometer was inserted into the water about three inches from the end. The exposed surface of the pipe (including the surface exposed by the thickness of the metal at the ends) was 287.177 square inches. The quantity of water contained in it was 132.534 cubic inches, and the equivalent to be added for the specific heat of the pipe was 39.341 cubic inches, making the estimated quantity of water 171.875 cubic inches." The temperature of the room in which the observations were conducted was 67° Fahr.

* I do not draw the same deductions from Mr. Tredgold's experiments that he does himself, and therefore did not give his figures here, but substituted my own in the manner just shown; the summary of the matter being that the heat lost through glass would be 2.248 heat-units, when that lost through iron would be 2.35 heat-units,

This pipe was presumably used on its sides in the horizontal position (though this is not stated), and represented no doubt a section of an ordinary 3-inch cast-iron heating-pipe, used at that time for green house heating, etc.

He informs us the rates of cooling were tried in different states of the surface. First, when in the usual state of cast-iron pipes covered with protoxide of iron (fine rust); second, black varnished; and third, with the varnish removed and two coats of white-lead paint substituted. He observed that the rusty surface cooled from 152° to 150° Fahr., or 2 degrees, in 2.5 minutes, and that it cooled from 150° to 140° Fahr., or 12 degrees, in fifteen minutes. This is at the rate of the whole quantity of water, or its equivalent, cooling one degree in 1.25 minutes.

He took observations every two degrees fall of the thermometer, which give slightly varying results as to the rate of cooling. This variation may be due to errors in reading the scales, or in errors in the thermometers, and a close study of the table of his experiments go to confirm the belief that for all practical purposes of house-warming the rate of cooling is very nearly directly as the temperature between pipe or or plate surface and the surrounding air. With the black surface of the pipe black-varnished, he found that to cool from 152° to 140° Fahr. (12 degrees) it took 14.533 minutes; or, in other words, cooled an average of one degree in 1.21 minutes, his readings showing a slight increase of cooling as the difference between surface and air became less. If we take the average of six experiments (1.23 minutes), progressing by two degrees, and correct the time observed on cooling the first two degrees by it, we have 2.42 minutes, instead of 2.266 minutes. This shows that the black-varnished surface is slightly more efficient than the rusty one—a little over three per cent.*

With the pipe with two coats of white-lead paint, the efficiency was less than with either of the others, but not as great as usually considered.

The cylinder cooled from 152° to 150° Fahr. (2 degrees) in (observed time) 2.316 minutes, and it cooled to 140° Fahr., or 12 degrees, in 15.366 minutes; or, in other words it cooled one degree in 1.28 minutes average.

Mr. Hood's summary of the matter is that 100 feet of varnished pipe, $103\frac{1}{4}$ feet of plain pipe and $105\frac{3}{4}$ feet of white-painted pipe have the same value as heating surfaces. He does not, however, give us the value of these surfaces in heat-units per square foot per degree difference, no more than Mr. Tredgold does, and as it is important, we shall have to calculate it for ourselves by the same method of reasoning, etc., as we did in the case of the latter's experiments.

The surface of the experimental piece of pipe is given as 287.177 square inches, which is two square feet, lacking less than one square inch, and therefore we will call it two square feet. The quantity of water actually contained in it was 132.534 cubic inches, and the equivalent in cubic inches of water that was to be added for the specific heat of the iron of the pipe, 39.341 cubic inches, making the estimated value of

---

* In comparing these statements with Mr. Hood's table, note that the time here is given in minutes and decimals of a minute, while in the table it is given in minutes and seconds.

the water and its envelope equal to 171.875 inches of water.

The water was cooled from 152° to 140° Fahr. in each experiment, and therefore had a mean temperature of 146° Fahr. The weight of a cubic inch of water at temperature is 248 grains; therefore we have

$$\frac{171.875 \text{ cub. in.} \times 248 \text{ grs.}}{7{,}000 \text{ grs. (1 lb.)}} = 6.089 \text{ lbs. of water.}$$

This water was cooled 12 degrees in the various times, which gives us $6.089 \times 12 = 73.068$ heat-units as the total heat given off in each from two square feet of heating surface, or 37.534 heat-units per square foot.

The air of the room was 67° Fahr., consequently the difference of temperature, or, in other words, the excess of the temperature of the surface over the air, was 89 degrees.

The time for cooling the rusty cylinder was 15 minutes, or one quarter of an hour; therefore we have

$$\frac{36.534 \times 4}{89° \text{ Fahr.}} = 1.642 \text{ heat-units}$$

per square foot per hour per degree difference. For the varnished surface it is 1.589 heat-units, and for the white-painted surface 1.552 heat-units.

To ascertain the effect of glass windows to cool the air of a room, Mr. Hood made experiments with a glass vessel as nearly as possible of the same thickness as ordinary window-glass. The temperature of the room was 65° Fahr., and the surface of the vessel was 34.296 square inches, and it contained 9.794 cubic inches of water, including the equivalent for the specific heat of glass. He does not tell us the form of the vessel, which would be very important to know, but presumbly, it was rectangular, or at least had perpendicular sides, and being small, represented an average effect in cooling, so that the deductions obtained are, presumbly, fully equal to average conditions.

The average rate of cooling from 150° to 110° Fahr. was found to be 1.176 degrees when the mean excess of temperature of surface was 65° Fahr. above the temperature of the air, and the time 34 minutes.

The total quantity of water, or its equivalent, is found to weigh .3482 pounds at a temperature of 130° (its mean temperature). This cooled 40° Fahr. = 13.93 heat-units for 34.296 square inches, or 58.48 heat-units for a square foot for 34 minutes, or 103.2 heat-units for an hour, divided by the mean difference in temperature $= \frac{103.2}{65°} = 1.59$ heat-units per square foot per hour per degree difference of temperature.

Mr. Hood's deductions from his experiment is to the effect that each square foot of window-glass will cool in a minute of the time 1.279 cubic feet of air as many degrees as the inside air is warmer than the external in a comparatively still atmosphere, but that whem windows are exposed to the action of winds further experiments are necessary.

It is evident the cooling of air through glass, etc., depends on both the velocity of the air inside and outside taken together.

Nearly all the heat that is lost by air of rooms to cooler air through glass is lost by convection. The air inside the glass falls by loss of heat and increase of weight and follows the laws of a falling body. The velocity of air outside is due to wind-pressure and the angle at which it strikes the glass. Quadrupling the velocity of the outer air, however, does not quadruple the loss of heat through the glass, for the reason that the air inside will not fall in the same ratio, but in a ratio about as the square root of the increase of outside velocity, so that the loss of heat through glass cannot be accurately established for a given difference of temperature and a certain velocity of the wind outside. An approximation, however, can be made to the loss of heat for other velocities and temperatures. Unfortunately, we have no very accurate data on the cooling effect of windows for the guidance of heating engineers, though on the warming effect of radiator surfaces there is not such a scarcity of information. Mr. George H. Barrus, of Boston, in experiments with a Walworth vertical wrought-iron pipe radiator for steam, found that under average conditions of use, with eight pounds of steam, in an atmosphere of about 51° Fah., that the units of heat given off per actual square foot of surface was 394.4. If we assume the surface of the iron to be 235° Fahr. (the temperature of the steam) we have 235° − 51° = 184° difference. Then

$$\frac{394.4}{184°} = 2.143 \text{ heat-units.}$$

This is somewhat less than Mr. Tredgold's experiments give for a short vertical cylinder, but it is what would be expected, as the pipes used were 30 inches long, and in a cluster, $2\frac{1}{4}$ inches between centres, screwed into a base.

He also experimented with a Nason radiator of ordinary height, two pipes wide by 24 pipes long.

The total number of heat-units per square foot of surface given off was 347.6, the pressure of the steam was eight pounds, and the temperature of the air of the room 64° Fahr. Assuming the temperature of the pipe to surface is 230° Fahr., and the difference then between air and heating surface is 170° Fahr. which gives us $\dfrac{347.6 \text{ heat-units}}{170°} = 2.045$ heat-units per hour per square foot per degree difference.

Mr. Barrus' method of measuring the heat was to receive the water of condensation carefully and to asertain its weight, then compute the heat according to the latent heat of steam. The nearness of the results thus obtained by vertical radiators of different makes, and at different times in different buildings, by the same methods, adds value to the data and establishes the fact, when taken with other investigations, that a tube of vertical radiator will give off heat equal to about *two heat-units per square foot per hour per degree difference.*

An experiment made by the writer in 1884 on a 2×7 Bundy steam-radiator, for his own information, and before the Bundy patterns were altered, to have an actual surface equal to their commercial rating, gave the following results:

Actual surface, 38 square feet; water condensed for one hour was 12.843 pounds, when the pressure of steam was maintained between 1 and 1½ pounds; temperature of air of room at floor commencement of experiment 52°; at 5 feet high on side wall 58°; temperature of air of room at floor at end of experiment 57½°; and at 5 feet 64°. The temperature of the air as it was found at the floor was, presumably, the temperature at which it first came in contact with the heater, but, as in other cases, the temperature of the room only was noted without informing us further, we will in this case take the mean of the temperature given, which is 57.9° Fahr., and, presumably, near enough for our purpose, which is not to compare rival heaters, but to establish the condensation or cooling for ordinary conditions of use.

Taking the temperature of the steam (one pound), therefore, at 215° Fahr., and the latent heat of its vaporization at 962 heat-units per pound, we will have —difference of temperature between steam (or pipe) 157.1° Fahr., and total heat of 12.843 pounds of steam, 12 355 heat-units, or 325.1 heat-units per actual square foot of surface, equaling 2.07 heat-units per square foot of surface per hour per degree of difference between steam and air.

It is possible that should these radiators be transposed as to the buildings they are tested in, the results would slightly differ, as the effect of the passage of heat by radiation alone from or to the radiator cannot be estimated, as it will depend on the surrounding walls, etc. For instance, one experiment being made in a cellar and another on the upper floor of a building, it is reasonable to assume the localities will effect results, and the question of humidity may also come in as a factor for or against a radiator. Draughts of air, also, will materially alter results, and the effect of an open hatchway, machinery in motion, or down draughts from windows, etc., will all tend to throw some uncertainty into the matter, so that unless positions, etc., are transposed and the water of condensation measured in the same manner and with similar apparatus, it would be difficult to determine positively which of the above radiators gives the highest result per actual square foot of surface. Such remarkable uniformity, however, by different makers appears to establish beyond a doubt that 2° Fahr. per square foot of surface per degree difference of temperature between the surface and air may be taken as the basis of loss of heat from vertical radiators, whether they are for hot water or steam.

This, of course, is the maximum, and it is for radiators of plain, smooth surface, say not over three feet in height, that are not covered up with screens or slabs, but used in the most practical manner and not too close to the walls. It should be borne in mind, also, that these radiators were only two pipes wide, and represented slightly more than the average, and that radiators of three or four pipes in width should not be expected to give quite as good results as radiators of one or two rows, other things being the same.

It is very possible, taking all the styles and kinds of radiators and coils known to the writer, that the minimum condensation or cooling may be placed at 1.25 heat-units and the maximum at 2 heat-units. Between these points it must be left to the judgment and experience of the fitter to select when the character of the radiator or coil is known.

There are certain extended surface direct radiators that will not give more than 1.25 heat-units; while long coils on the walls of, say, one-inch pipe, and not too high—say six pipes high—will probably run up to 2.25 heat-units.

Professors Denton and Jacobus, of the Stevens Institute of Technology, found that an extended surface vertical sectional radiator gave 0.0017 pounds of condensation per degree difference of temperature per hour per square foot of average surface, with steam at 3.9 pounds pressure; which is approximately 1.63 heat-units, while the same radiator gave 0.00204 pounds of condensation per actual remaining square foot of surface, the extended surface having been planed off, which is approximately 1.95 heat-units. This confirms my opinions given many years ago in the earlier editions of this book. It does not mean, however, that the radiator, as a whole, did more work. It simply means that a unit of the remaining surface was about 20 per cent. more efficient, and that generally an equal amount of iron in a plain direct radiator is very much more efficient than in an extended surface radiator.

I wish, also, to add that I am not convinced of the general inferiority of extended surface radiators for indirect work. The value unit for unit of surface, as between extended and plain surface indirect radiators, I will admit, is in favor of the plain surface; but when it comes to a matter of dollars and cents, or pounds weight of iron used, I believe the extended surface inditect radiator to be the most economical. This comes from the inability to arrange round iron pipes or flat surfaces so that the air passages will be properly proportioned to the least amount of surface used, to get the required results. In other words, the cast-iron extended surface can be arranged so that the air can be brought in more intimate contact with it.

Returning again to condensation in direct radiators, the writer here introduces the results of some of his own experiments made with cast-iron sectional radiators. The results do not show as high a rate of condensation as those given previously; but this may be chiefly attributed to the fact that the experiments were made in cotton-covered boxes or frames, with only six inches of space between the sides of the radiators and the cotton-covered frames. Top and bottom they were open, but in these confined spaces they could not do the work they would in a large room under conditions similar to their constant use. For comparison, however, they were probably all that could be required. I may add here that under conditions said to be about equal to those of actual use, Professors Denton and Jacobus have reported condensation equal to 2.4 heat-units for one type of direct radiator, but which forms entirely too high an average for ordinary use.

The following I take from a paper I contributed to the American Society of Civil Engineers, read at the June convention of 1894; and I introduce it here more to confirm the fact that the line of difference of temperature between the radiator surface and that of the temperature of the air of the room agrees very nearly with the condensation line in a good radiator—so near, in fact, as to be practically the same.

The writer had occasion recently, in connection with his professional work, to make some comparisons as to the relative efficiencies of several of the regular makes of sectional cast-iron radiators now on the market, and ones that are considered good types of direct cast-iron radiators. The results of the experiments hereafter given were obtained from four of these radiators that were selected as a standard of comparison for a new radiator about to be put on the market. The results of these experiments are given to this Society, as they form valuable data by which an engineer can approximate the amount of condensation likely to take place in the heating apparatus of a building, and also show how near the line of condensation agrees in steam pipes with a line of the difference of temperature between·air and steam.

The writer is compelled to omit the name and make of these four types, but as the results run so nearly together he presumes it is not necessary to go further than to say that they are the ordinary well-known types of cast-iron sectional radiators that are so much used at the present day. All radiators were of 48 sq. ft. of surface, according to their commercial rating, excepting that designated as No. 3, which contains 52 sq. ft., and which, when reduced to the value of 48 sq. ft., gives the dotted line shown and designated as $X-X$, (Diagram Fig. 122).

The method of comparison used was that known as the condensation test, the only one which is of any value in determining the comparative value of radiators, and consisted of taking steam from a boiler at 90 lbs. pressure and conveying it in a 1-in. pipe to a water separator. This water separator was "trapped" in the usual manner with a Nason trap. From the top of the separator a 1-in. pipe led to a reducing pressure regulating valve, and from the pressure regulating valve it led into a header of 2 ins. inside diameter. The connection for each radiator was made of $\frac{1}{2}$-in. pipe, taken from the upper side of this header through a short nipple, and thence by another short nipple into the radiator, an angle valve being used at the corner. It is probable that the condensation taking place in the two short nipples and in the valve was carried forward

DATA OF CONDENSATION IN 'RADIATORS.

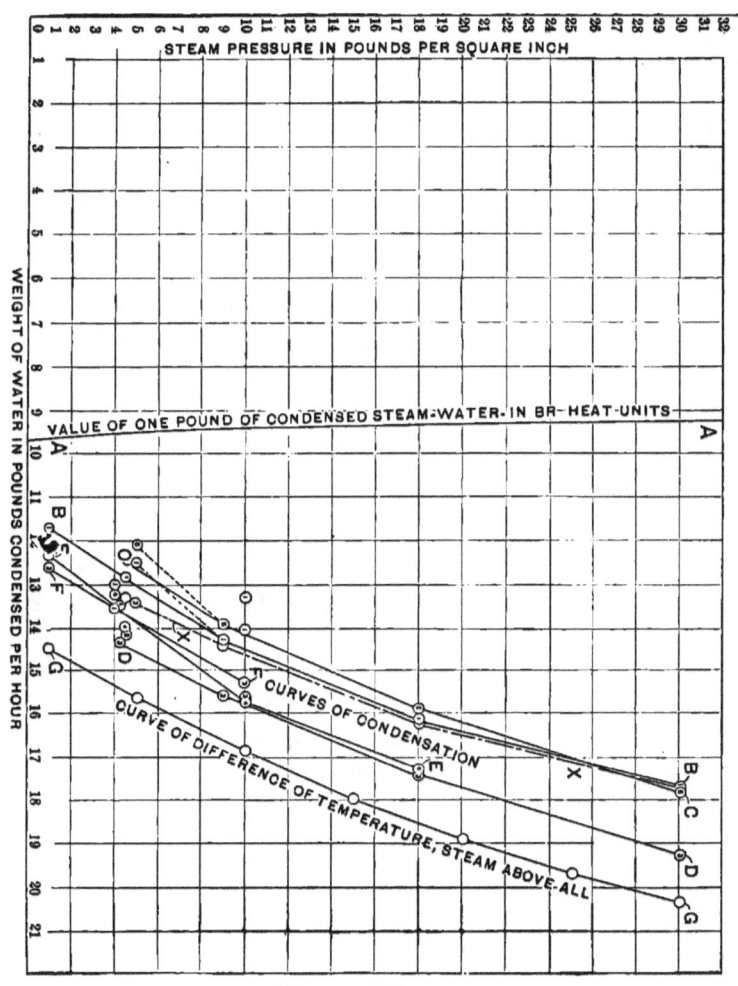

Diagram, Fig. 122.

into the radiator; but, the quantity being so small, it has not been considered, as it was the same in each case.

The radiators were placed on platforms 2 ft. above the floor and 36 ins. from center to center. Over each radiator was placed a frame covered with muslin. The muslin in all cases was 6 ins. from the sides and ends of the radiators, and the frames were opened top and bottom. The frames had the same perpendicular height as the radiator, and were raised sufficiently (about 6 ins.) to permit an uninterrupted flow of air to the radiator underneath the lower edge of the frame. This method enclosed each radiator in a cotton box, as it were, and interposed two thicknesses of muslin between opposing sides of radiators, so as to cut off almost entirely the effects of radiation from one radiator to another. The tables on which the radiators were set had an inclination of about 1 in. to the drip end, so that the condensation could easily gravitate to the apparatus for collecting the water.

The apparatus for collecting and measuring the water consisted of a $\frac{3}{8}$-in. "tail pipe," on the side of which was a short ordinary water-gauge glass. Both above and below the water glass in the tail pipe was a stop valve, the upper valve being for the purpose of shutting off the connection should the gauge glass break; but during the experiments it was always open. The lower valve was simply a tail valve, to control the escape of water. The tail valve was supplied with 1 ft. of very light rubber hose ($\frac{3}{8}$ in.) which terminated in a paper pail. The water was drawn into the paper pail through the tail valve, but was always kept in sight in the glass and made to agree with a thread around the middle of the glass, both at the commencement and the end of all trials.

On the header before mentioned was arranged a steam-pressure gauge, and in an oil well in the upper side of the same header was fixed a standard thermometer for the purpose of checking errors in the steam-pressure gauge by comparisons with the temperature of the steam. It will be noted that the trials were generally made at 1, 5, 10, 18 and 30 lbs. pressures, but that some of the circles indicating trials are on other lines, notably in the neighborhood of 5 and 10 lbs. Where the greater number of trials are shown these differences of pressure are the results of corrections made by the thermometer, the pressure gauge being somewhat inaccurate and the pressures likely to fluctuate $\frac{1}{2}$ lb., or thereabouts, as a pressure-regulating valve is only an approximate instrument for work of this kind. At the lower pressures (1 to 5 lbs.) there is very little loss due to the evaporation of the hot water as it falls into the pails, especially as the end of the hose is kept under water as much as possible. At the higher trials a small percentage of the hot water, when relieved of its pressure, is converted into steam. This is obviated in a great measure, however, by keeping the end of the hose always under water, which can be done as soon as condensation forms in the bottom of the pail and covers the mouth of the hose.

It is well to remark here that the header before spoken of, and in which steam at reduced pressures was admitted to the radiators, was also drained and kept dry by an automatic steam trap. The positions of the radiators were also changed from time to time, as it was feared that a radiator in one position might be favored.

At the left of the diagram is shown the steam pressure in pounds per square inch, at which various trials were made.

The next line, A—A, gives the value in British heat-units of each pound weight of steam for the various pressures, and is measured from the left-hand side of the diagram, each of the larger squares having a value of 100 heat-units, the thermal value of a pound weight of steam being less from the higher pressures.

The curved line B—B shows the results obtained in the radiator designated as No. 1, and they were made under conditions of very nearly equal temperature, in 73° and 74° Fahr.

Line C—C shows results obtained for radiator marked No. 6 in the trials, which were made at the same pressures and the same temperatures as the foregoing.

Line D—D gives the curve of condensation in the radiator marked No. 3, temperatures and pressures of steam being the same as before.

Line E—E shows the results of experiments on radiator No. 4, which at the lower temperatures seems not to do so well. Between the 10-lb. and 18-lb. lines, however, the curve agrees with the remaining curves of condensation.

Line F—F shows the other experiments made with radiator No. 3, when the temperature of the room averaged 77° and 78°. This high temperature of the room undoubtedly accounts for the line showing less condensation on that day. The circles 6, 1 and 3 on the 5-lb. line are also results of the tests made when the temperature of the room was about 78°. Other circles, enclosing figures in close proximity to lines, with corresponding figures, show values obtained with the same radiator during other trials and under slightly different conditions of the temperature and humidity. They all harmonize, however, with the curves of their respective radiators and tend to confirm the results.

Line G—G is the curve of difference of temperature between the steam and the air, and is drawn for comparison with the curves of condensation as discovered by the trials. The uniformity that exists is remarkable. There is a correction, however, to be made for the higher temperatures in the matter of the curves of condensation. As the temperature of the steam advances, the value of 1 lb. of condensation becomes less, as shown in the line A—A, so that curves of efficiency of condensation, based on the actual amount of heat lost from the radiator, and not from the number of pounds of water condensed, would all tend a little to the left as they go upward, bending away from the curve of difference of temperature. It is well known that the condensation does not increase directly as the difference of temperature. These experiments, however, show that it (the difference) is not as great as was surmised, and that for all practical purposes the curve of condensation and the curve of difference of temperature is practically alike for any ranges of temperature and pressures carried in low-pressure steam-heating apparatus.

Since the foregoing paper was written, I have had

time to have an assistant tabulate the results of the experiments given above, which are as follows:

TABLE.

AVERAGE OF RESULTS OBTAINED WITH 3 RADIATORS OF 48 SQ. FT. EACH.

Steam-pressure per Square Inch.	Steam Condensed by a 48 Square Foot Radiator per Hour.	Latent Heat of Steam, B. H. Units.	Total Heat in B. H. Units.	Square Foot of Surface in Radiator.	Units of Heat per Square Foot per Hour.	Difference of Temp. in Degrees Fahr. between Steam and Air.	Units of Heat per Sq. Ft. per Hour per One Degree Difference of Temperature.
1	12	962	11544	48 sq. ft.	240.5	141.5	1.70
5	13	953	12389	48 "	258.1	153.5	1.68
10	14.4	945	13608	48 "	283.5	165.5	1.71
15	15.5	938	14539	48 "	302.9	176.5	1.71
20	16.3	932	15191.6	48 "	316.5	185.5	1.70
25	17.1	926	15834.6	48 "	329.9	194.0	1.70
30	17.8	921	16393.8	48 "	341.5	200.5	1.70
I.	II.	III.	IV.	V.	VI.	VII.	VIII.

RESULTS OBTAINED WITH 1 RADIATOR OF 52 SQ. FT.

1	13.6	962	13083.2	52 sq. ft.	251.6	141.5	1.77
5	14.6	953	13913.8	52 "	267.5	153.5	1.74
10	15.8	945	14931	52 "	287.1	165.5	1.73
15	16.8	938	15758.4	52 "	303	176.5	1.71
20	17.7	932	16496.4	52 "	317.2	185.5	1.70
25	18.6	926	17223.6	52 "	331.2	194.0	1.70
30	19.3	921	17775.3	52 "	341.8	200.5	1.70

CHAPTER XXX.

PIPE COVERING—WHAT IS SAVED THEREBY, AND OTHER DATA.

In a work of this kind I cannot go into the relative merits of pipe covering. There are so many things to be considered and so many interests involved that it would prove of no special benefit to the steamfitter, and would raise endless discussion, with which no one would be satisfied. The subject may be treated of generally, however, with a great deal of benefit to all those concerned, and I will therefore give some general deductions, obtained in 1892 by the author while making experiments for the H. W. Johns Mfg. Co. of New York. I was assisted by the well-known steam engineer, Mr. George H. Barrus, of Boston, whose name is a sufficient guarantee of accuracy and integrity in trials of this kind.

The company mentioned were desirous of comparing not only their own special products relatively, but also of comparing them with other well-known coverings then on the market.

It is needless to go into the method used, other than to say that they were condensation tests, and that the utmost precaution was taken to prevent inaccuracy. Many experiments and repetitions of the experiments were made, until the methods used developed a uniformity of results which were in themselves a check on all possible mistakes of manipulation or observation that were likely to creep in.

Experiments were always made on a two-inch pipe about 16 feet long, and having exactly 10 square feet of surface to a pipe. These pipes were called "elements" of 10 square feet each, and twelve of them were arranged in twelve separate chambers, open top and bottom, all within a very large loft of a factory where the temperature was nearly constant, and so arranged that one could not influence the other. Transposition was also resorted to to check inequalities due to neighboring influences.

The pressures at which the trials were made were 50, 100, 150 and 200 pounds, and presumably the trials represented the most elaborate experiments ever completed in this direction, and it will be a gratification to the writer could he make the results public, but as they are the property of the company who paid for them, this cannot be done. Certain deductions, however, can be made public, and these I give, with the consent of the parties most interested.

The diagram, Fig. 123, is taken from the report. The horizontal lines show pressures of steam in pounds above atmosphere. The vertical lines show pounds weight of steam condensed per element of 10 square feet of pipe surface per hour. The line AB is the line of condensation within one of the elements without covering. It shows the condensation in an uncovered two-inch pipe. The lines marked 1 to 12, inclusive

PIPE COVERING. 321

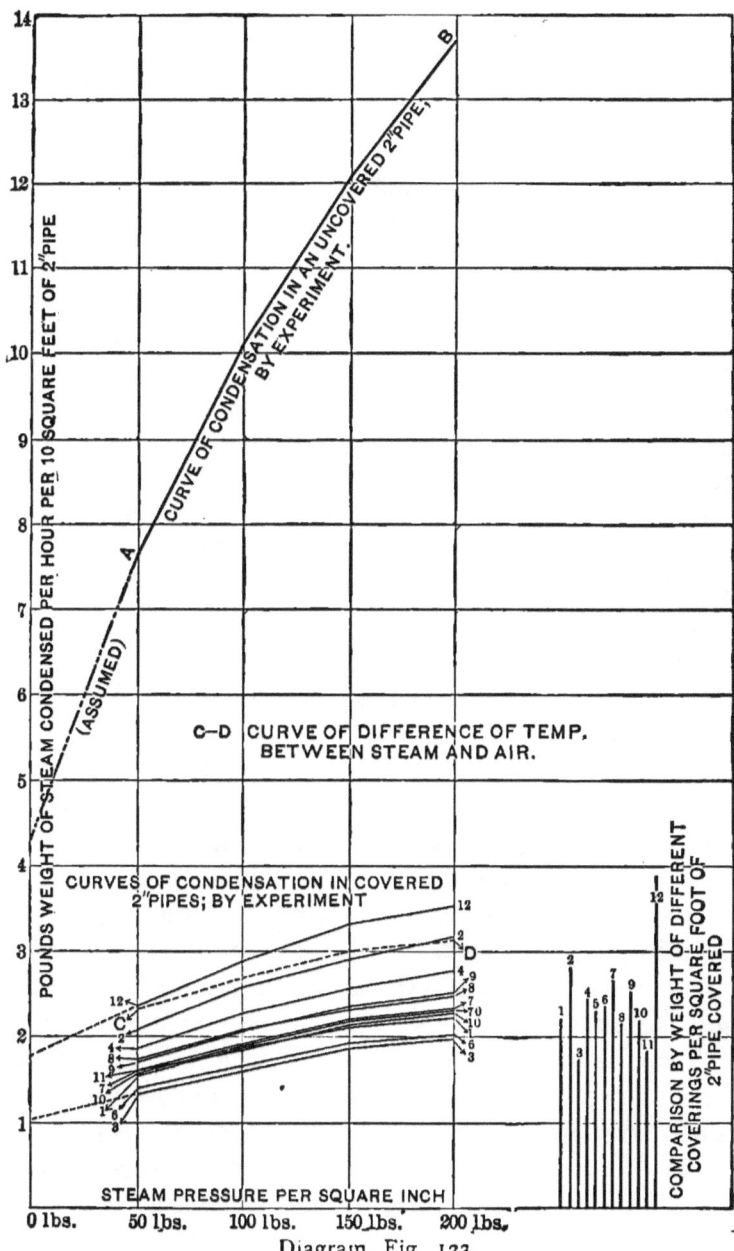

Diagram, Fig. 123.

show the amount of condensation for each element for twelve selected commercial pipe coverings. The line CD is an arbitrary line, conforming to the differences of temperature between that of the air and that of the steam within the pipe, and shows a remarkable uniformity with lines of condensation. The vertical lines at the right show the comparative weights of the different coverings per square foot of pipe surface covered.

An important deduction to be drawn is that the weight of the covering bears a close analogy to the efficiency of the covering; generally the lighter the covering, the better it is. It will be noticed that covering No. 3 gave the least condensation, and consequently the highest efficiency; and it will be seen by the vertical lines that its weight was the least. It will also be noticed that covering No. 12, the poorest covering and the one that gave the highest condensation, weighs the most, and is about double the weight of No. 3, while the condensation is also approximately nearly twice as great.

Another and startling deduction to be drawn from the chart is that the very poorest covering (No. 12) is so immensely better than no covering, that where it is necessary to save steam or prevent heating or warming a room or space, the covering of steam pipes is certainly one of the best of investments. I may add, "cover them with anything" rather than leave them uncovered.

The distance from the curved line AB to the base line shows the condensation in an uncovered pipe. The distance from the other curves to the base line gives the condensation according to the covering used. It will be noticed that even with the poorest covering, that from $\frac{2}{3}$ to $\frac{3}{4}$ of the total condensation due to an uncovered pipe can be prevented, while with the best of coverings between $\frac{5}{6}$ and $\frac{6}{7}$ of the total condensation will be prevented. This is the result that went beyond the author's expectation, and will, no doubt, astonish many others, now that it is pointed out; as, in my judgment, few had a full appreciation of the matter.

In regard to the relative efficiency of the different coverings, it will be noticed that the greatest condensation for 10 square feet of surface is 3 pounds 7 ounces of water in an hour's time (No. 12), when the pressure of steam was 200 pounds to the square inch, while at 50 pounds pressure of steam it was 2 pounds $7\frac{1}{2}$ ounces. This was with the poorest covering, while at the same time the condensation in the uncovered pipe was over $13\frac{1}{2}$ pounds at 200 pounds pressure, and nearly $7\frac{3}{4}$ pounds at 50 pounds pressure. It will further be noticed that the best covering reduced the condensation for a pressure of 200 pounds to 2 pounds $15\frac{1}{2}$ ounces, while at 50 pounds it was but 1 pound 7 ounces, showing that between a poor and a good covering there was a very material difference. In other words, the best covering was nearly twice as efficient as the poorest.

The various intermediate results for other conditions can be approximated from the diagram. It was explained that the line AB was the line of condensation in an uncovered pipe. The length of the vertical lines intersecting the curved line gives the condensation for the different pressures. In like manner the vertical distance from the base line to the ordinates of curves gives the condensation for 10 square feet of pipe surface, covered with the different materials (1 to 12).

The accompanying table shows the weight of steam condensed per hour for each square foot of surface. Column No. 1 shows the number of the pipe or element, and refers to the covering used. Columns Nos. 2, 3, 4 and 5 show the condensation per square foot of pipe covered for the various pressures of 50, 100, 150 and 200 pounds of steam above temperature. Element No. 5 indicates the condensation in the uncovered pipe.

TABLE NO. 8.

POUNDS OF STEAM CONDENSED PER HOUR PER SQUARE FOOT OF 2' PIPE.

No. of Pipe.	50 Lbs.	100 Lbs.	150 Lbs.	200 Lbs.
1	.1549	.1854	.2135	.2232
2	.2091	.2591	.2945	.3178
3	.135	.1602	.1872	.197
4	.1878	.2292	.2574	.2778
5*	.7682	1.013	.....	.....
6	.1408	.1681	.194	.2007
7	.1599	.1909	.2194	.2324
8	.1745	.2069	.2324	.2497
9	.1703	.2059	.2359	.2532
10	.1605	.1862	.2132	.2284
11	.1629	.1911	.2209	.2331
12	.2352	.2885	.3325	.3537
Average of all but No. 5........	.1719	.2065	.2364	.2515

* No. 5 shows the condensation in an uncovered pipe.

There are other conditions to be considered in selecting coverings than efficiency. A covering must be permanent, and not easily charred or deteriorated by high-pressure steam; and I mention it here to prevent a person from selecting a covering by its weight alone.

CHAPTER XXXI.

MISCELLANEOUS NOTES.

CUTTING WALLS AND COVERING RISERS.

ARCHITECTS often omit to leave a recess where required in the walls of a building, and the fitter has to cut one. In his anxiety to put up as much pipe as possible, and as he considers the cutting of the wall does not properly belong to him, he cuts it in the quickest and easiest manner he can, regardless of the appearance, and in some loosely put up walls it is a difficult task to make an attractive or even satisfactory piece of work. The proper way would be to have the openings left and cutting avoided; but if it must be done, it should be well done.

Let the fitter provide himself with sharp chisels and a light hammer, and he can generally cut a brick without disturbing it in the wall; but it is also necessary for the master mechanic to consider wall-cutting labor, and to give the workman to understand he will be credited with cutting walls as well as for putting up pipe.

The fitter should get the architect's permission before he commences to cut, for otherwise there may be much injury done to a building by having a recess cut from top to bottom near a front wall or corner, or where considerable weight comes on the wall.

The best way to cover a riser recess is with a board lined with tin on its inner side. Have the grounds put on before the plastering is done, and have the panel screwed on afterward. The panel may also be fancy iron-work, with holes in it, which makes a very permanent method. A moveable panel admits of access to the pipes to make repairs without breaking the walls.

Some architects require the recess to be plastered over, using slate or coarse wire-cloth to hold the plaster, so as to entirely hide the appearance of a pipe, but even then they do not entirely succeed, for two or three reasons. When a slate is stuck over the recess with plaster-of-paris, and plastered over all, the expansion of the slate often cracks the plaster. When plastered on wire-cloth, it does very well, and will not crack, but it will turn a dark color in time, as will any thin covering when it becomes warm, because the continuous current of air passing up the wall at that particular spot deposits more dust there than at any other point, and leaves a well-defined mark.

For the same reason, the walls back of radiators get dark more rapidly than the walls of any other part of the room. The same is true of curtains which hang near a register. In parts of the country where soft coal is generally used this is very apparent.

TURNING EXHAUST STEAM OR VAPOR INTO CHIMNEYS.

There is a custom among steam-fitters and others of turning the exhaust from an engine into the boiler chimney in buildings, ostensibly to "make the draft better," but in reality to save running an exhaust pipe to the roof of the building. Exhaust steam, turned into a long or high brick chimney, will not improve the draft, but impair it.

In locomotives the exhaust steam is turned into the stack to increase the draft, and in short iron stacks of portable engines it has the same effect, when properly put in ; but it must be borne in mind, that to be effective, it must have such proportions as to make it an injector, to increase the velocity of the air by contact with its own high velocity, before it has time to expand and fill the stack.

In long iron stacks, a little steam turned into them may, perhaps, be of some use in warming the stack (which cools rapidly from contact with the wind and air in cold weather), and by assisting the upward current of smoke or air by mixing with it. Under certain conditions, it makes a mixture of steam and air lighter than the air alone. If the increased velocity caused thereby more than compensates for the extra volume which has to pass, it may possibly be an improvement.

But usually the exhaust steam impairs a long chimney (especially an iron one), leaving the condensation to run down the inside of the chimney in streams, and to be again partly re-evaporated by absorbing heat from the gases of combustion. In brick chimneys this is very apparent, condensing and soaking into the brick-work, and absorbing as much heat from the gases of combustion to re-evaporate and drive off a cubic foot of it as would cool 30,000 cubic feet of air 100 degrees Fahrenheit. It also destroys the chimney.

SOLDERING OF PIPES AND BRASS FITTINGS.

Often it is necessary to solder or "sweat" pipes into fittings, or male and female threads of brass work. The latter is no trouble, and can be done by tinning the parts to be put together, using only resin for a flux, if done while new, and then screwing them together while hot. When iron pipe has to be sweated into iron fittings, malleable iron fittings should be used, because they can be tinned by using muriatic acid reduced with zinc; cast-iron does not solder well.

When about to sweat a pipe and fitting together, wipe the threads carefully, and run a carefully-wiped die over the male thread, to entirely clean it, using a clean tap to remove any oxide or grease from the female thread in the fitting; then tin cleanly, using muriatic acid for a flux, and screw the parts together while both are hot.

The assertion is sometimes made that solder will melt under high-pressure steam temperatures. This I found to be so when the pressure of the steam went beyond 180 pounds per square inch. The solder used in "sweating" brass pipes together was what is known as "half and half"—that is, about one-half tin and one-half solder, no bismuth being used that I could discover. When the steam pressure was between 180 and 190 pounds, or at a temperature of about 380° Fahr., the solder oozed out of the threads like drops of quicksilver, and I found it necessary to make the threads tight by screwing alone. Of course this temperature of 380° or 382° Fahr. is considerably below the melting point even of tin (about 444° Fahr.), but it was probably sufficient to soften the alloy so that it flowed under the pressure of 180 pounds per square inch.

There is no advantage in soldering a frost burst in an iron pipe, through which steam or very hot water passes, for it will not last.

In iron water-pipe, rather than remove the pipe, it may be soldered, but it must be thoroughly cleaned and tinned, and a heavy wipe joint made on it; bolting is of no avail.

When cracks appear in brass or copper pipes without any apparent cause, there is very little use in soldering, for they are usually caused by undue expansion in adjacent parts of the metal, and are a fault of manufacture which soft solder will remedy for a very short time only.

PAINTING PIPES.

Distributing pipes may be painted with anything that will arrest oxidation. Raw linseed oil, with ochre of the required color, and turpentine, form a good preparation for radiators, when they are to be bronzed, as it gathers and "fixes" any machine oil or dirt there may be on the pipes, and will make a good back for the bronze. When radiators are painted steam should be turned on immediately, and they should be dried by heat. If they are painted with heavy oil paints that dry on the surface and are allowed to stand for some time, when steam is turned on they will blister.

White or colored enameled paints make a good finish for radiators, but it is difficult to obtain a paint that will not change its color under heat. Dead or flat paints, with plenty of turpentine, should be used first, and the finishing coat should be the enameled paint. For hot water work enameled paints stand very well.

Black baking japan, or black air-drying japan, are very good substances for painting pipes and iron-work, and two coats will give a good gloss, which does not require to be renewed. A wipe with a slightly oiled woolen cloth will give them a fresh appearance. Bronzing cannot be done over a black varnish, which will show through many coats of bronze.

Black paraffine varnish should not be used, as it is not permanent; it cokes with heat, and has no body.

Indirect coils, or coils or heaters which cannot be seen, it is best not to paint.

Dust allowed to collect on heaters impairs their efficiency very much.

CHAPTER XXXII.

FIRE FROM STEAM PIPES.

THERE is much diversity of opinion as to whether wood, in its *simple* state, painted or unpainted, or its charcoal, will take fire from the heat of steam, *unsuperheated*, at any ordinary pressures. Steam can be made so hot by superheating that it will ignite wood, and its pressure may be made so high its heat will be sufficient to ignite wood; but these are not ordinary conditions of steam any more than they would be ordinary conditions of air, and air also can be warmed either by superheating or pressure until it will ignite wood. What evidence, therefore, have we that steam at ordinary conditions will ignite wood or other woody fibres?

I have known tampico fibre that was used in the manufacture of brushes to fall between the heating pipes of the drying room, in which there was almost constantly 60 pounds pressure of steam, and although it was as fine as bristles, and was packed in between pipes, it did not take fire. I have also used pine laths on the upper side of the shell of a horizontal boiler, to maintain a space, when turning the arch of brickwork over the boiler, and in years afterwards, when resetting the same boilers, I found these laths in good condition and not charred, although they were subject to the heat of 60 pounds pressure of steam and in heavy and close contact with the iron of the boiler, with 4 inches of brickwork as a covering to the laths, which, of course, prevented any great loss of heat through or from the outer surface of the wood.

The forging and the lagging of engine cylinders with wood furnish apparently good evidence against the ignition of wood by steam pipes at ordinary low pressures and at the ordinary condition of steam.

Superheated steam, however, and very high-pressure steam can undoubtedly ignite wood. I have seen the hair, felt, and canvas burned off the pipes of a heating apparatus for 20 or 30 feet from the boiler by the water becoming low or the boiler nearly empty and the steam becoming superheated.

In all construction, therefore, care should be taken to guard against the contact of the steam pipes with wood, as the general danger is undoubtedly increased by the omission to do so.

Some charcoals will ignite at a much lower temperature than others, and it is a well-known fact that the lower the temperature at which charring occurs, the lower the temperature of ignition. The question is, however, whether the temperature of charring can ever become so low as to cause the temperature of ignition to become equally low, or nearly as low.

The question has been discussed in the technical schools and societies and by the insurance companies without any definite conclusion. I think, however, it is well to perpetuate a diagram (Fig. 124) that appeared in the *Scientific American* about twenty years ago, which "Mr. Stahl, a student of the graduating class of the Stevens Institute of Technology, prepared, at the request of Professor Thurston, in which the vertical scale is one of temperatures of preparation of charcoal, and the horizontal scale is one of temperatures of ignition, and the curve shown contains the points of correspondence as given in the table.

"It will be seen that the curve is apparently nearly hyperbolic. The lowest temperature of preparation was 500° Fahr., but it is seen at a glance, even that at 350°, the temperature of steam under a pressure of over 125 lbs. per square inch, the temperature of preparation and of ignition cannot coincide unless some marked change of law should occur at so low a temperature, carrying the curve, which here represents that law, abruptly inward to reach the point A. It is needless to state that such a phenomenon would be quite improbable, and is probably impossible."

The foregoing was the editorial comment that appeared with the diagram.

When the writer first saw the above he was induced to make some crude experiments in the same direction.

The pine laths that I before mentioned I enclosed in a retort, and to prove this wood was not charcoal, I placed it in a retort and drove off gas that burned with nearly as much light as illuminating gas when it leaves the retort.

I inclosed a two-inch cube of white pine wood within a small gas pipe retort, with a bit of solder (one-third tin and two-thirds lead) and a bit of sheet lead, and placed the retort in a boiler tube for five days, boiler going day and night. At the end of that time the wood was pure charcoal, the solder was melted, and the lead was not, which goes to show pure charcoal can be made at a temperature between 440° and 612° Fahr.; it being understood that the melting point of this solder is given at 441° Fahr. and that of the lead 612° Fahr.

To prove the above was pure charcoal, *i. e.*, that all the hydrocarbon was driven off, I raised the temperature of the retort to about 1,200°, but could not drive off any more gas.

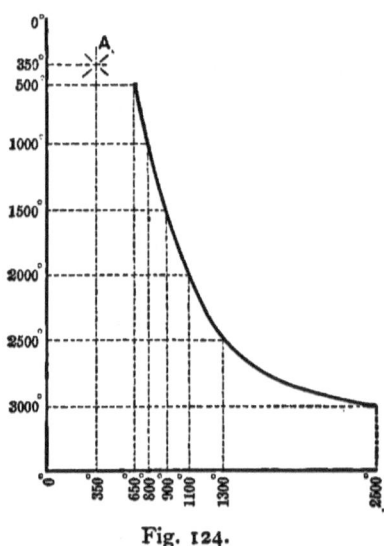

Fig. 124.

I then inclosed pine laths against the shell of a horizontal boiler, and covered them with a course of brick on edge. The pressure of steam in this boiler was from 40 to 60 pounds day and night for about 2½ years, except one day a month for cleaning. The ends of the laths that came out to the air and flush with the brickwork were not near as dark as hemlock tanned leather, and the darkest part I could find which was entirely covered with brick was not as dark as roasted coffee. This goes to show charcoal cannot be made at 300° Fahr., after $2\frac{1}{2}$ years, under the most favorable circumstances, with a furnace fire only 5 feet beneath it.

In experiments on the ignition of charcoal, I found that the charcoal made in a boiler tube would not redden at the melting point of lead (612° Fahr.), but would at a lower temperature than zinc (770° Fahr.).

My mode of operation was this way. I passed a gas pipe through a fire and blew pure hot air through the pipe. I also prepared myself with long, slender strips of solder (half and half, and one-third tin and two-thirds lead), and with strips of lead and zinc, and pine shavings, and small pieces of the laths and charcoal.

The pure charcoal would not redden in the same blast that just melted the lead, but did in a blast which melted it rapidly. When held in a blast which melted solder (one-third tin and two-thirds lead, melting temperature about 500° Fahr.), it showed no signs of fire or redness.

The lath, which was $2\frac{1}{2}$ years in contact with the boiler under a course of brick, would become charcoal in a temperature which melted half and half solder, but would not get a spark on it until I increased the temperature to where the needle of lead bent and dropped. It was the same with a nicely prepared splinter of white pine, and I could see no deviation in the action from the splinter of the lath; they all became charred in the blast which melted half and half solder, but would not take on a spark until the lead melted.

With a blast that fused a metal 19 parts tin, 31 lead and 50 bismuth, melting temperature about 212° Fahr., I could not turn tissue paper brown.

Gunpowder held in the blast which melted the lead did not explode until after the lead melted. It gave off a slight blue sulphurous light first, then the lead melted, and an instant after the powder exploded.

Illuminating gas will not take fire from a cherry red poker, but will from a bright red one.

The gas of wood, crude petroleum, soft coal, or any other hydrocarbon, will not take fire when escaping hot from the retort. With a cherry red poker I have tried the three mentioned.

CHAPTER XXXIII.

MISCELLANEOUS NOTES AND TABLES.

These notes and tables will be found of service in estimating.

The avoirdupois pound is always to be used, unless otherwise specified. It contains 7,000 Troy grains; the grain is always Troy.

16 drams	= 1 ounce.	oz.
16 ounces	= 1 pound.	lb.
25 pounds	= 1 quarter.	qr.
4 quarters	= 1 hundred.	cwt.
20 cwt., 2,000 lbs.	= 1 ton.	

The gross ton (in which the quarter becomes 28 lbs., the hundredweight, 112 lbs., and the ton, 2,240 lbs.) is used in estimating English goods at the U. S. Custom-House; in freighting; in the wholesale coal trade; and in the wholesale iron and plaster trades, and when specified.

$$1 \text{ lb. avoir.} = 16 \text{ oz. avoir.} = 7,000 \text{ grs. Troy.}$$
$$1 \text{ ``} \quad \text{``} = 437.5 \text{ ``} \quad \text{``}$$

$27\frac{7}{10}$ cubic inches of water weigh one pound avoirdupois, at a temperature of 40°.

A cubic foot of water, at a temperature of 60°, weighs 999 ozs., and is taken as 1,000 ounces, or 62½ pounds, for all ordinary calculations. It weighs a little less than 60 pounds when the temperature is 212°.

A cubic foot of water contains very nearly 7½ gallons, and for rough calculations may be taken as such (7.4805 gallons is actual) number.

A cubic inch of water, at its greatest density, weighs 252.725 grains; a cubic foot, 62.4 pounds.

			1 gal. =	231.0 cubic in.
	1 cub. ft.,	7½ " =	1728.0	"
1 bushel, 1 $\frac{6}{25}$	"	9$\frac{3}{10}$ " =	2150.42	"
1 cord,	128.0	" " =		"
1 cub. yd.,	27.0	" " =	46656.0	"
1 barrel,*	4.21	" 31½ " =	7276.5	"

* A flour barrel will hold 33.28 gallons, or 4.449 cubic feet, or 2.79 heaped bushels (called 2¾ bushels).

In estimating quantities of water by barrels, 31½ standard gallons equals the barrel.

TABLE No. 9.

WEIGHT OF A CUBIC INCH OF VARIOUS METALS.

Iron, cast	0.263 of a pound.
" wrought	0.28 "
Lead	0.41 "
Copper	0.32 "
Nickel	0.30 "
Steel	0.28 "
Tin	0.265 "
Zinc, cast	0.24 "
" rolled	0.26 "
Brass, steam metal	0.315 "
" yellow	0.282 "

TABLE NO. 10.

WEIGHT OF A CUBIC FOOT OF VARIOUS BUILDING MATERIALS, IN POUNDS (APPROXIMATE).

Granite	168.0 pounds
Marble	165.0 "
Sandstone	135.0 "
Blue-stone	165.0 "
Slate	180.0 "
Mortar, dry	80.0 to 100 pounds.
Common Brick	112.0 pounds.
Dry Sand	100.0 "
Fire-brick	135.0 "

One perch of stone-work, in walls or foundations, measures 24¾ cubic feet.

One thousand common bricks, laid in a wall, makes about 50 cubic feet, varying a little for different bricks.

Six fire-bricks to each square foot of lining, one brick thick, is sufficient; 1,000 bricks will make 170 superficial feet of lining, laid in the ordinary way.

To find the weight of iron castings by computation.—Find its solid contents, in inches, and multiply them by 0.26, and it will give the weight, in pounds. For rough calculations, it will do to divide the cubic inches by 4, and call the answer pounds.

To find the weight of any other casting, or forging.—Find its solid contents in cubic inches, and multiply by the weight of a cubic inch of the metal, as given in the table No. 9, "Weight of a cubic inch of various metals."

For irregular castings, which are difficult to measure, and cannot be conveniently weighed, a *rough estimate* of their weight may be taken, provided they are not *cored out*, by weighing the pattern, if it is of soft pine, and allowing 13 times the weight of the pattern, if it is new, or just out of the sand, and 14 times if it has laid in the pattern loft for some time.

A square foot of cast-iron, one inch thick, weighs 37½ pounds. To find what a square foot of any other thickness will weigh, multiply 37½ by the thickness in inches, or fractions of an inch.

A square foot of rolled wrought-iron, one inch thick, weighs 40 lbs. To find the weight of boiler plates, or sheet-iron, per square foot, multiply 40 by the decimal of an inch in thickness the required plates are to be.

TABLE No. 11.

THE FOLLOWING TABLE SHOWS THE DIFFERENCE BETWEEN AMERICAN AND ENGLISH WIRE GAUGES, AND THE THICKNESS OF PLATES, IN DECIMALS OF AN INCH FOR EACH.

No. of Gauge.	American. Inches.	English. Inches.
0000	0.46	0.454
000	0.4096	0.425
00	0.3648	0.38
0	0.3248	0.34
1	0.2893	0.3
2	0.2576	0.284
3	0.2294	0.259
4	0.2043	0.238
5	0.1819	0.22
6	0.1620	0.203
7	0.1442	0.18
8	0.1284	0.165
9	0.1144	0.148
10	0.1018	0.134
11	0.0907	0.12
12	0.0808	0.109
13	0.0719	0.095
14	0.0640	0.083
15	0.057	0.072
16	0.05	0.065
17	0.045	0.058
18	0.04	0.049
19	0.035	0.042
20	0.031	0.035

To find the weight of a cast-iron pipe, for one foot of its length.—Multiply the diameter of the pipe in inches by 3.1416, and multiply the answer thus obtained by the thickness of the pipe in inches, or decimals of an inch, then by 12 and 0.26 respectively; or, instead of the last two, use 3.15.

This will give about the weight of the pipe, including the hubs, as the outside circumference of the pipe is not the mean length of the iron, according to its thickness. To be exact. Proceed as above, but take one thickness of the iron from the diameter of the pipe first, and it will give the weight of the pipe without hubs or flanges.

Example.—Required the weight of a 12-inch pipe, ½ inch thick, for one foot of its length. Thus : 12. in.—0.5 =11.5 × 3.1416 = 36.127 × 0.5 =18.063 × 3.15=56.89 pounds.

The 3.15 is the product of 12 inches for the length and 0.263 for the weight.

Definitions and computations in mensuration, required by the steam-fitter.

The perimeter of a figure is its outer boundary, without regard to shape.

A true circle forms the shortest perimeter for the greatest area inclosed, and is called a *circumference*.

A diameter is a right line, passing through the center of a circle.

A diameter is very nearly $\frac{32}{100}$ of the circumference of the same circle, or, to be exact, 0.3183 of it. Rule.—Multiply the circumference by 0.3183, and it will give the answer, in the same denomination.

A circumference is $3\frac{14}{100}$ of the diameter of the same circle very nearly, or, to be exact, 3.1416.

The square of the diameter of a circle is multiplying it once by itself. Thus, if the diameter is 4, the square will be 16. (4 inches × 4 inches = 16 inches.)

To find the area (the number of square inches) within a circle.—Multiply the square of the diameter by 0.7854, and it will give the answer in the same denomination as it was squared in. Thus, $4'' \times 4'' = 16'' \times 0.7854 = 12.566$ square inches, whose diameter is 4 inches.

The cube of a number is the number multiplied by itself twice. Thus, $4 \times 4 = 16 \times 4 = 64$.

When the cube of the diameter of a sphere is multiplied by 0.5236, it gives the solid contents, in numbers of the same denomination as it was cubed in. Thus: $4'' \times 4'' = 16'' \times 4'' = 64'' \times 0.5236 = 33.51$ cubic inches, for a ball four inches in diameter; and when multiplied again by 0.263 it gives 8.813, which will be the weight in pounds of a cast-iron ball of the same diameter.

A cylinder of the same length as its diameter has the same surface as a sphere of equal diameter (surface of ends, of course, not included).

To find the surface of a cylinder 4 inches in diameter and 4 inches long.—Multiply the diameter by 3.1416 and the product by the 4 inches in length. Thus, $4 \times 3.1416 = 12.566 \times 4 = 50.2656$, the square inches on the outside of a 4 × 4 cylinder.

To find the surface of a sphere 4 inches in diameter. —Square the diameter, and multiply by 3.1416. Thus: $4 \times 4 = 16 \times 3.1416 = 50.2656$.

To find the outside surface of a pipe.—Multiply the outside diameter in inches by 3.1416, and by the length in inches, and divide by 144; it will give the answer in square feet.

To find the pressure, per square inch, a column of water of any height will exert.—Multiply the height of the column, in feet, by the weight of a cubic foot of water in pounds at the temperature the water may be, and divide by 144.

Example.—Required the pressure, per square inch, of a head of water of 200 feet, and when the temperature of the water is 40° Fahr. (weight 62½ pounds). Thus, 200 × 62.5 = 12500 ÷ 144 = 86.8 pounds per square inch.

Required the pressure of the water at a temperature of 212°. Thus, 200 × 59.80 = 1196 ÷ 144 = 83.05 pounds per square inch.

TABLE No. 12.

THE FOLLOWING TABLE OF DIAMETERS, CIRCUMFERENCES, AND AREAS IS GIVEN FOR "READY-RECKONING."

Diameter.	Circumference.	Area.	Diameter.	Circumference.	Area.
$\frac{1}{16}$	0.1963	0.0030	$1\frac{7}{16}$	4.5160	1.6229
$\frac{1}{8}$	0.3927	0.0122	$1\frac{1}{2}$	4.7124	1.7671
$\frac{3}{16}$	0.5890	0.0276	$1\frac{9}{16}$	4.9087	1.9175
$\frac{1}{4}$	0.7854	0.0490	$1\frac{5}{8}$	5.1051	2.0739
$\frac{5}{16}$	0.9817	0.0767	$1\frac{11}{16}$	5.3015	2.2365
$\frac{3}{8}$	1.1781	0.1104	$1\frac{3}{4}$	5.4978	2.4052
$\frac{7}{16}$	1.3744	0.1503	$1\frac{13}{16}$	5.6941	2.5801
$\frac{1}{2}$	1.5708	0.1963	$1\frac{7}{8}$	5.8905	2.7611
$\frac{9}{16}$	1.7671	0.2485	$1\frac{15}{16}$	6.0868	2.9483
$\frac{5}{8}$	1.9635	0.3068			
$1\frac{1}{16}$	2.1598	0.3712	2 in.	6.2832	3.1416
$\frac{3}{4}$	2.3562	0.4417	$\frac{1}{16}$	6.4795	3.3411
$1\frac{3}{16}$	2.5525	0.5185	$\frac{1}{8}$	6.6759	3.5465
$\frac{7}{8}$	2.7489	0.6013	$\frac{3}{16}$	6.8722	3.7582
$1\frac{5}{16}$	2.9452	0.6903	$\frac{1}{4}$	7.0686	3.9760
			$\frac{5}{16}$	7.2649	4.2001
1 in.	3.1416	0.7854	$\frac{3}{8}$	7.4613	4.4302
$\frac{1}{16}$	3.3379	0.8861	$\frac{7}{16}$	7.6576	4.6664
$\frac{1}{8}$	3.5343	0.9940	$\frac{1}{2}$	7.8540	4.9087
$\frac{3}{16}$	3.7306	1.1075	$\frac{9}{16}$	8.0503	5.1573
$\frac{1}{4}$	3.9270	1.2271	$\frac{5}{8}$	8.2467	5.4119
$\frac{5}{16}$	4.1233	1.3529	$1\frac{1}{16}$	8.4430	5.6727
$\frac{3}{8}$	4.3197	1.4848	$\frac{3}{4}$	8.6394	5.9395

Diameter.	Circumference.	Area.	Diameter.	Circumference.	Area.
8 3/16	25.7218	52.8994	3/4	36.9138	108.4342
8 1/4	25.9182	53.4562	7/8	37.3065	110.7536
8 5/16	26.1145	54.2748			
8 3/8	26.3109	55.0885	12 in.	37.6992	113.0976
8 7/16	26.5072	55.9138	1/8	38.0919	115.4660
8 1/2	26.7036	56.7451	1/4	38.4846	117.8590
8 9/16	26.8999	57.5887	3/8	38.8773	120.2766
8 5/8	27.0963	58.4264	1/2	39.2700	122.7187
8 11/16	27.2926	59.7762	5/8	38.6627	125.1854
8 3/4	27.4890	60.1321	3/4	40.0554	127.6765
8 13/16	27.6853	60.9943	7/8	40.4481	130.1923
8 7/8	27.8817	61.8625			
8 15/16	28.0780	62.7369	13 in.	40.8408	132.7326
			1/8	41.2338	135.2974
9 in.	28.2744	63.6174	1/4	41.6262	137.8867
1/16	28.4707	64.5041	3/8	42.0180	140.5007
1/8	28.6671	65.3968	1/2	42.4116	143.1391
3/16	28.8634	66.2957	5/8	42.8044	145.8021
1/4	29.0598	67.2007	3/4	43.1970	148.4896
5/16	29.2561	68.1120	7/8	43.5857	151.2017
3/8	29.4525	69.0293			
7/16	29.6488	69.9528	14 in.	43.9824	153.9384
1/2	29.8452	70.8883	1/8	44.3751	156.6995
9/16	30.0415	71.8121	1/4	44.7676	159.4852
5/8	30.2379	72.7509	3/8	45.1605	162.2956
11/16	30.4342	73.7079	1/2	45.5532	165.1303
3/4	30.6306	74.6620	5/8	45.9459	167.9896
13/16	30.8269	75.6223	3/4	46.3386	170.8735
7/8	31.0233	76.5887	7/8	46.7313	173.7820
15/16	31.2196	77.5613			
			15 in.	47.1240	176.7150
10 in.	31.4160	78.5400	1/8	47.5167	179.6725
1/8	31.8087	80.5157	1/4	47.9094	182.6545
1/4	32.2014	82.5160	3/8	48.3021	185.6612
3/8	32.5941	84.5409	1/2	48.6948	188.6923
1/2	32.9868	86.5903	5/8	49.0875	191.7480
5/8	33.3795	88.6643	3/4	49.4802	194.8282
3/4	33.7722	90.7627	7/8	49.8729	197.9330
7/8	34.1649	92.8858			
			16 in.	50.2656	201.0624
11 in.	34.5576	95.0334	1/8	50.6583	204.2162
1/8	34.9503	97.2053	1/4	51.0510	207.3946
1/4	35.3430	99.4121	3/8	51.4447	210.5976
3/8	35.7357	101.6234	1/2	51.8364	213.8251
1/2	36.1284	103.8691	5/8	52.2291	217.0772
5/8	36.5211	106.1394	3/4	52.6218	220.3537

MISCELLANEOUS NOTES AND TABLES. 345

Diameter.	Circumference.	Area.	Diameter.	Circumference.	Area.
2 13/16	8.8357	6.2126	5 1/2	17.2788	23.7583
2 7/8	9.0321	6.4918	5 9/16	17.4751	24.3014
2 15/16	9.2284	6.7772	5 5/8	17.6715	24.8505
			5 11/16	17.8678	25.4058
3 in.	9.4248	7.0686	5 3/4	18.0642	25.9672
3 1/16	9.6211	7.3662	5 13/16	18.2605	26.5348
3 1/8	9.8175	7.6699	5 7/8	18.4569	27.1085
3 3/16	10.0138	7.9798	5 15/16	18.6532	27.6884
3 1/4	10.2102	8.2957			
3 5/16	10.4065	8.6179	6 in.	18.8496	28.2744
3 3/8	10.6029	8.9462	6 1/16	19.0459	28.8665
3 7/16	10.7992	9.2806	6 1/8	19.2423	29.4647
3 1/2	10.9956	9.6211	6 3/16	19.4386	30.0798
3 9/16	11.1919	9.9678	6 1/4	19.6350	30.6796
3 5/8	11.3883	10.3206	6 5/16	19.8313	31.2964
3 11/16	11.5846	10.6796	6 3/8	20.0277	31.9192
3 3/4	11.7810	11.0446	6 7/16	20.2240	32.5481
3 13/16	11.9773	11.4159	6 1/2	20.4204	33.1831
3 7/8	12.1737	11.7932	6 9/16	20.6167	33.8244
3 15/16	12.3700	12.1768	6 5/8	20.8131	34.4717
			6 11/16	21.0094	35.1252
4 in.	12.5664	12.5664	6 3/4	21.2058	35.7847
4 1/16	12.7627	12.9622	6 13/16	21.4021	36.4505
4 1/8	12.9591	13.3640	6 7/8	21.5985	37.1224
4 3/16	13.1554	13.7721	6 15/16	21.7948	37.8005
4 1/4	13.3518	14.1862			
4 5/16	13.5481	14.6066	7 in.	21.9912	39.4846
4 3/8	13.7445	15.0331	7 1/16	22.1875	39.1749
4 7/16	13.9408	15.4657	7 1/8	22.3889	39.8713
4 1/2	14.1372	15.9043	7 3/16	22.5802	40.5469
4 9/16	14.3335	16.3492	7 1/4	22.7766	41.2825
4 5/8	14.5299	16.8001	7 5/16	22.9729	41.9974
4 11/16	14.7262	17.2573	7 3/8	23.1693	42.7184
4 3/4	14.9226	17.7205	7 7/16	23.3656	43.4455
4 13/16	15.1189	18.1900	7 1/2	23.5620	44.1787
4 7/8	15.3153	18.6655	7 9/16	23.7583	44.9181
4 15/16	15.5716	19.1472	7 5/8	23.9547	45.6636
			7 11/16	24.1510	46.4153
5 in.	15.7080	19.6350	7 3/4	24.3474	47.1730
5 1/16	15.9043	20.1290	7 13/16	24.5437	47.9370
5 1/8	16.1007	20.6290	7 7/8	24.7401	48.7070
5 3/16	16.2970	21.1252	7 15/16	24.9354	49.4833
5 1/4	16.4934	21.6475			
5 5/16	16.6897	22.1661	8 in.	25.1328	50.2656
5 3/8	16.8861	22.6907	8 1/16	25.3291	51.0541
5 7/16	17.0824	23.2215	8 1/8	25.5255	51.8486

Diameter.	Circumference.	Area.	Diameter.	Circumference.	Area.
7/8	53.0145	223.6549	7/8	67.1517	358.8419
				67.5444	363.0511
17 in.	53.4072	226.9806		67.9371	367.2849
1/8	53.7999	230.3308		68.3298	371.5432
1/4	54.1926	233.7055	7/8	68.7225	375.8261
3/8	54.5853	237.1049			
1/2	54.9780	240.5287	22 in.	69.1152	380.1336
5/8	55.3707	243.9771	1/8	69.5079	384.4655
3/4	55.7634	247.4500	1/4	69.9006	388.8220
7/8	56.1561	250.9475	3/8	70.2933	393.2031
			1/2	70.6860	397.6087
18 in.	56.5488	254.4696	5/8	71.0787	402.0388
1/8	56.9415	258.0161	3/4	71.4714	406.4935
1/4	57.3342	261.5872	7/8	71.8641	410.9728
3/8	57.7269	265.1829			
1/2	58.1196	268.8031	23 in.	72.2568	415.4766
5/8	58.5123	272.4470	1/8	72.6495	420.0049
3/4	58.9056	276.1171	1/4	73.0422	424.5577
7/8	59.2977	279.8110	3/8	72.4349	429.1352
			1/2	73.8276	433.7371
19 in.	59.6904	283.5294	5/8	74.2203	438.3636
1/8	60.0831	287.2723	3/4	74.6130	443.0146
1/4	60.4758	291.0397	7/8	75.0057	447.6992
3/8	60.8685	294.8312			
1/2	61.2612	298.6483	24 in.	75.3984	452.3904
5/8	61.6539	302.4894	1/8	75.7911	457.1150
3/4	62.0466	306.3550	1/4	76.1838	461.8642
7/8	62.4393	310.2452	3/8	76.5765	466.6380
			1/2	76.9692	471.4363
20 in.	62.8320	314.1600	5/8	77.3619	476.2592
1/8	63.2247	318.0992	3/4	77.7546	481.1065
1/4	63.6174	322.0630	7/8	78.1473	485.9785
3/8	64.0101	326.0514			
1/2	64.4028	330.0643	25 in.	78.5400	490.8750
5/8	64.7955	334.1018	1/8	78.9327	495.7960
3/4	65.1882	338.1637	1/4	79.3254	500.7415
7/8	65.5809	342.2503	3/8	79.7181	505.7117
			1/2	80.1108	510.7063
21 in.	65.7936	346.3614	5/8	80.5035	515.7255
1/8	66.3663	350.4970	3/4	80.8962	520.7692
1/4	66.7590	354.6571	7/8	81.2889	525.8375

To find the circumferences of *larger* circles, multiply the diameter by 3.1416. For areas, multiply the square of the diameter by 0.7854.

TABLE NO. 13.

SHOWING THE NUMBER OF FEET IN LENGTH OF VARIOUS SIZED PIPES WHICH WILL CONTAIN ONE CUBIC FOOT OF WATER.

Nominal Size of Pipe.	Length in feet which will contain one cubic foot.	Nominal Size of Pipe.	Length in feet which will contain one cubic foot.
$\frac{1}{2}$	470.0	$3\frac{1}{2}$	14.6
$\frac{3}{4}$	270.0	4	11.3
1	167.0	$4\frac{1}{2}$	9.
$1\frac{1}{4}$	96.5	5	7.2
$1\frac{1}{2}$	70.5	6	5.
2	43.9	7	3.54
$2\frac{1}{2}$	30.0	8	2.875
3	19.35	9	2.26

By multiplying the above lengths by the relative volume * of steam at any required pressure, it will give the length of pipe which will be necessary to contain a cubic foot of water when converted into steam at that pressure. This is necessary in ascertaining the amount of water that will be taken from a boiler to fill the pipes and radiators of the apparatus with steam.

To make the subject of horizontal multi-tubular boilers complete for the class of men who do not find it convenient or cannot spare the time to figure the matter out for themselves, or who may not have the data at their fingers' ends, the annexed table (No. 14) and diagram have been worked out to cover the principal points connected with this class of boilers.

The diagram Fig. 125 of the head-sheets shows the number of tubes of the ordinary sizes in general use that can be properly and conveniently put into the head-sheets of boilers from 36 inches to 60 inches in diameter, at the same time leaving sufficient room

* See Table No. 5, page 177.

above the tube line to form a steam space and allow sufficient distance for a safe water line.

The diameters of boilers and tubes given cover the common range of sizes used in warming. An empirical rule exists among some boiler constructors to have one-fourth of the diameter of the boiler above the tube line to allow for steam space and water above the tubes. This may do on boilers of large diameter, but for 36 and 42 inch boilers, practice has proved that it is not sufficient—at least in boilers that are to be used for warming purposes with gravity apparatus. When a boiler has to supply steam for engines and power steadily, and has a continuous and regular feed supply that will keep the water within a range of rise and fall of two inches, then the minimum water and steam space of $\frac{1}{4}$ can be used as a good common rule, provided tube surface enough cannot be properly placed in less than the remaining three-fourths of the boiler heads. But this one-fourth rule should not be followed in boilers under 48 inches in diameter, even for power and seldom in boilers for heating purposes, or for heating purposes and power combined.

The reason for this is well known to the experienced steam-heating engineer, and is obvious when it is explained that a boiler or a battery of boilers must have sufficient water above the tube lines to allow for the filling of all the mains and heater with steam without lowering the water to a dangerous point.

A novice about a steam-heating boiler will have an experience about as follows: He will get up steam and fill his mains and radiators to, say five pounds pressure. As the water lowers in his boiler he adds fresh water to maintain the water line, say "two

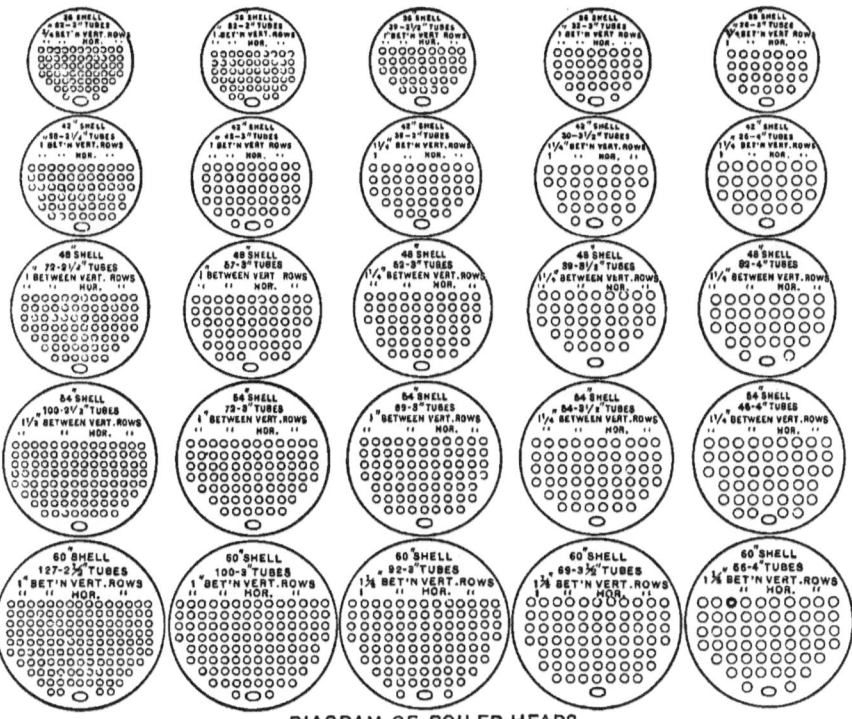

DIAGRAM OF BOILER-HEADS

Fig. 125.   [*To face page* 348.]

cocks." As soon as the water of condensation begins to come back he finds the water will rise in his boiler, and it puzzles him a little. However, he shortly realizes it is the return water, and goes to the blow-off cock and lets some of it out. Soon he gets his heating apparatus in "train," and the water "stands" all right, and his first scare is over. Then he receives an order to "run the pressure up" to 50 pounds. Away goes the water again; in fact, it has been lowering in the glass since the pressure began to increase. Caution drives him to the pump, and he puts the water up to "two cocks" again, and all goes well for a time. Then he receives an order to open the doors and let the pressure down, and from that moment he is in trouble again, and it may be he will find he has water enough coming from some place to fill the boiler to the safety-valve, and he goes again to the blow-off cock.

Now, all this is wrong. There should be sufficient water above the tubes of such boilers to fill the mains and radiators with steam at the highest pressure likely to be carried, without having to add more, unless some has been wasted by blowing into the sewer or elsewhere.

If the filling of the mains and radiators of a steam apparatus with steam at five pounds pressure takes $1\frac{1}{2}$ inches of water from a boiler, it will take $4\frac{1}{2}$ inches if the pressure is run up to 50 pounds, and to this should be added $2\frac{1}{2}$ inches for safety, making the level of the water at the starting of the fire seven inches or more above the tube line.

Exactly what this distance should be cannot be accurately established, and will be different in different apparatus, increasing with the bulk of steam within the radiators, as compared to the surface of the radiators.

For the above reason another empirical rule was given for a general determination of the water line and the steam space above the water, and good practice had established this to be about two-fifths of the diameter of the boiler.

On the other hand, again, we are generally confronted with the necessity of getting the largest possible tube surface in a given size shell, on account of the restricted width and height of the space allowed for boilers in most of our large buildings. As a compromise between the two empirical rules the boiler heads shown in the diagram have been laid out in every case to give about one-third of the diameter of the boiler above the tube line, and to be as near the two-fifths line as is consistent with getting a reasonable tube area, both in regard to heating surface and flue area.

The rule has been to keep all tubes three inches from the shell of the boiler, as called for in the laws governing marine boilers, and in these diagrams no tube comes nearer the shell than $2\frac{1}{4}$ inches.

There are two instances, also, in which a greater distance can be obtained between the tube line and the top of the boiler-shell than shown in the diagrams, and that is in the case of the 36-inch boiler, with thirty-nine $2\frac{1}{2}$-inch tubes, and the 48-inch boiler, with thirty-nine $3\frac{1}{2}$-inch tubes. In both cases the tubes could be dropped $1\frac{1}{2}$ inches without bringing them too close to the shells.

It appears unnecessary to explain the object of Table

No. 14; it shows for itself. It was made by the author for his own convenience, and its use must be apparent to any one who wishes to know how many tubes he can put into a certain diameter boiler without resorting to the drawing board to lay it out for himself. Thus he finds the number in column 4, and the size of the tubes in column 2. Then he finds the heating surface of the boiler for given lengths in square feet in columns 8 and 9, but it does not include the upper half of the shell, which should not be counted. If, then, he wants the surface for other lengths of boiler he finds the number of square feet of surface he is to add or take away for each foot the boiler is longer or shorter than for the columns 8 and 9. The columns 10 and 11 give the tube area, the first in square inches and the latter in square feet. Columns 18 and 19 give the horse-power, assuming 15 square feet of average shell and tube surface to be one nominal horse-power in this class of boilers.

In the diagram, where tubes are dotted they can be dropped to good advantage, as being either too near the shell or in the way of inspection.

The thickness and areas of tubes used are those given by the National Tube-Works.

Diameter of Boiler-shells in Inches.	External Diameter of Tube in Inches.	Distance between Tubes in Vertical Rows.	Distance between Tubes in Horizontal Rows.	Total Number of Tubes in Boiler.	Length of Tube—Internal Diameter—to One Square Foot of Surface.	Square Feet of Heating-surface to One Foot Length of Boiler.	Square Feet of Heating-surface for Boiler 12 ft. Long.	Square Feet of Heating-surface for Boiler 16 ft. Long.	Total Cross-section of Internal Area of Tubes in Square Inches.
36	2	¾	¾	62	2.110	34.09	416.16	....	159.526
36	2	1	1	52	2.110	29.35	359.26	....	133.796
36	2½	1	1	39	1.674	28.00	343.06	....	159 510
36	3	1	1	32	1.370	28.06	343.78	....	194.240
36	3	1¼	1	26	1.370	23.68	291.22	....	157.820
36	4	1¼	1	16	1.024	20.33	251.06	....	175.024
42	2½	1	1	58	1.674	40.14	491.31	....	237.220
42	3	1	1	45	1.370	38.34	469.68	....	273.150
42	3	1¼	1	39	1.370	33.96	417.17	....	236.730
42	3½	1¼	1	30	1.172	31.03	382.62	....	250.410
42	4	1¼	1	26	1.024	30.88	380.22	....	284.414
48	2½	1	1	72	1.674	49.29	604.06	801.20	294.48
48	3	1	1	57	1.370	47.88	587.19	778.64	345.99
48	3	1¼	1	52	1.370	44.24	543.32	720.24	315.64
48	3½	1¼	1	39	1.172	39.55	487.16	645.36	325.533
48	4	1¼	1	32	1.024	37.53	462.92	613.04	350.048
54	2½	1	1	100	1.674	66.80	....	1084.67	409.00
54	3	1	1	72	1.370	59.62	....	969.79	437.04
54	3	1¼	1	69	1.370	57.43	....	934.75	418.83
54	3½	1¼	1	54	1.172	53.14	....	866.11	450.738
54	4	1¼	1	45	1.024	51.01	....	832.03	492.255
60	2½	1	1	127	1.674	83.71	....	1359.05	519.43
60	3	1	1	100	1.370	80.84	....	1313.13	607.00
60	3	1¼	1	92	1.370	75.00	....	1219.69	558.44
60	3½	1¼	1	69	1.172	66.72	....	1087.21	575.94
60	4	1¼	1	56	1.024	62.53	....	1020.17	612.58
1	2	3	4	5	6	7	8	9	10

TABLE NO. 14.

MISCELLANEOUS NOTES AND TABLES.

Total Cross-section of Internal Area of Tubes in Square Feet.	Internal Area of One Tube in Square Inches.	Distance between Tube-line and Top of Shell.	Thickness Nearest Birmingham Wire-gauge.	Thickness of Tube in Inches, National Tube Works.	Length of Tube—External Diameter—to One Square Foot of Surface.	Square Inch of Metal in Cross-section of One Tube.	Nominal Horse-power for Boiler 12 ft. Long.	Nominal Horse-power for Boiler 16 ft. Long.
1.107	2.573	13	13	.095	1.900	.560	27.75	....
0.929	2.573	14	13	.095	1.900	.560	23.95	....
1.107	4.090	13	12	.109	1.500	.819	22.86	....
1.343	6.079	13	12	.109	1.273	.990	22.91	....
1.095	6.079	15	12	.109	1.273	.990	19.41	....
1.215	10.939	14½	10	.134	.955	1.627	16.74	....
1.647	4.090	16½	12	.109	1.500	.819	32.75	....
1.896	6.079	16	12	.109	1.273	.990	31.31	....
1.643	6.079	16½	12	.109	1.273	.990	27.81	....
1.737	8.347	17	11	.120	1.090	1.274	25.57	....
1.974	10.939	16	10	.134	.955	1.627	25.34	....
2.045	4.090	19	12	.109	1.500	.819	40.27	53.41
2.402	6.079	19	12	.109	1.273	.990	39.14	51.90
2.191	6.079	19	12	.109	1.273	.990	36.22	48.01
2.260	8.347	18	11	.120	1.090	1.274	32.57	43.02
2.430	10.939	19	10	.134	.955	1.627	30.86	40.86
2.840	4.090	20	12	.109	1.500	.819	....	72.31
3.035	6.079	20	12	.109	1.273	.990	....	64.65
2.908	6.079	20	12	.109	1.273	.990	....	62.31
3.130	8.347	20	11	.120	1.090	1.274	....	57.74
3.418	10.939	20	10	.134	.955	1.627	....	55.46
3.607	4.090	21	12	.109	1.500	.819	....	90.60
4.215	6.079	21	12	.109	1.273	.990	....	87.54
3.878	6.079	21	12	.109	1.273	.990	....	81.31
4.000	8.347	21	11	.120	1.090	1.274	....	72.48
(Table 14)	10.939	21	10	.134	.955	1.627	....	68.01
11	12	13	14	15	16	17	18	19

TABLE NO. 14.—(*Continued.*)

INDEX.

CHAPTER I.
GRAVITY-CIRCULATING APPARATUS.

	PAGE
Gravity Systems of Piping	1
Nomenclature	6
Water-line	8
How a Building is Piped	10
Two Heaters from the Same Connection	10
Outlets of the Risers	11
Risers	11
Radiator Connections	12
Steam-mains (see Chapter XV)	13
Return of the Water under all Conditions of Pressure	14
The Size of Mains	15
How Steam-pipes should leave the Boiler	16
Relief Pipes	16
Pitch of the Main	16
Tees in a Main	17
Stop-valves in Risers	19
Stop-valves in Mains	19
Main Return-pipes	20
Dry Return-pipes	21
Check-valves in Returns	22
Circulating Pipe	23

CHAPTER II.
RADIATORS AND HEATING SURFACES.

Vertical Tube Radiators	24
Cast-iron Sectional Radiators	25

	PAGE
Steam entering Radiator	26
Description of Plate I	27
Plane Surfaces (Definition)	29
Extended Surfaces (Definition)	29
Bundy Radiator	30
Reed Cast-iron Loop Radiator	31
Box Coil	31
Gold Pin Radiator (Indirect)	32
Gold's Compound Coil Surface	33
Coils	34

CHAPTER III.

CLASSES OF RADIATION.

How Direct-radiating Surfaces should be placed	37
Indirect Radiators	38
Indirect-radiator Boxes	39
Air-flues	40
Change of Air in Rooms	41
Direct-indirect Radiation	42
Position for Indirect Heaters with the Action of Air in Rooms, etc.	45
Cold-air Inlet-ducts	46
Switch-valve and Arrangement of Indirect Surface for Schools.	49

CHAPTER IV.

HEATING SURFACES OF BOILERS.

Fire-box and Flues	50
Crowding the Fire-box with Hanging Surfaces	52
Corrugated Fire Surfaces	53
Boilers which have given the Best Results	53
Proportioning Boilers	54
Can a Boiler be robbed of its Heat by the Gases of Combustion?	54
Reverberatory or Drop-flue Boilers	55
Will the Quantity of Water within a Boiler effect Evaporation?	56

CHAPTER V.

BOILERS FOR HOUSE HEATING.

Simplicity of Parts	57
Requirements for House Boilers	57

	PAGE
Construction of Upright Boilers	59
Construction of Horizontal Boilers	60
Contracted Passages under Boilers	60
Technical Names of Parts of Boilers, and their Setting	61

CHAPTER VI.

FORMS OF BOILERS USED IN HEATING.

A Source of Danger to the Fitter	64
Upright Boiler without Tubes	65
Upright Multi-tubular Boiler	66
Upright with Steam-dome	67
Upright Drop-tube Boiler	68
Base-burning Boiler	72
Mills Cast-iron Sectional	73a
Mercer Cast-iron Sectional	73a
Gurney Cast-iron Sectional	73b
Cottage	75
Doric	76
Bundy	77
Horizontal Tubular Boilers	78
Horizontal Multi-tubular Boilers	81
Water-tube Boiler	83
Root Water-tube Safety Boiler	84

CHAPTER VII.

GENERAL REMARKS ON BOILER SETTING.

Thickness of Walls	85
Marshy or Sandy Ground	85
Why Boiler Walls crack	85
Fire-bricks in a Furnace	87
Front-connection Division	88
Dead Plates	89
Bridge-walls	89
Ash-pits	89
Lugs on Boilers	90
Suspending Boilers from I Beams	91

CHAPTER VIII.

PROPORTION OF THE HEATING SURFACES OF BOILERS TO THE HEATING SURFACES OF BUILDINGS.

	PAGE
Relation of Boiler to Heating Surface of a Building	93

CHAPTER IX.

GRATES AND CHIMNEYS.

Grate of a House Boiler	99
Size of Grate to Boiler	100
Evaporation per Pound of Coal	100
Air-space in Grates	102
Size of Chimneys	103
Examples of Grates and Chimneys	105
Testing Efficiency of Chimney by Water Gauge	108
Theoretical Intensity of Chimney	109
Table of Grates and Chimneys	111
Why Grates break	112

CHAPTER X.

SAFETY-VALVES.

Boilers bursting when working at Ordinary Pressures	114
The Office of the Safety-valve	115
Decrease of Pressure under the Valve	116
Table of Lift of a 4-inch Valve at Various Pressures	117
Graphic Illustration of the Size of the Opening of a 4-inch Valve when blowing off at Various Pressures	117
Formulæ for calculating the Size of Safety-valves	118
Construction and Operation of Safety-valves	121
Ordinary Safety-valve with Auxiliary Attachment	122
Water-column Safety-valve	123
Safety-valve with Pipe carried below Water-line	125
Pop Safety-valves	125
To calculate Weight necessary to retain given Pressure	126

CHAPTER XI.

DRAFT REGULATORS.

	PAGE
Diaphragms	129
Construction of Regulators	130
High-pressure Draft Regulator	131
Connecting Regulators	131
How Regulators are attached to Ash-pit Doors	133
Setting Doors for Regulators	133

CHAPTER XII.

AUTOMATIC WATER-FEEDERS.

Construction	134
When a Water-feeder should be used	137
Connections to Water-feeders	138
Draught in Pipes	138
Fluctuations of Water in a Gauge Glass	139

CHAPTER XIII.

AIR-VALVES ON RADIATORS.

Where they should be placed	140
Drawing Air from Coils, etc.	141
Old-style Air-valves	143
Modern Air-valves	145

CHAPTER XIV.

STEAM PIPE, SIZE, AREA, EXPANSION, ETC.

Description of Pipe	148
Nominal Size of Pipe	148
Table of Standard Dimensions of Pipes	149
How to calculate the Relative Areas of Pipes	150
Table of Relative Areas of Pipes	153
Diagram of Relative Areas of Pipes	155

	PAGE
Expansion of Pipe and its Relation to Steam-mains	156
Expansion of Return-pipes	157
Effect of Lime and Moisture on Pipes	157
Expansion of Pipes buried in the Ground	158
Expansion-joints and how to Compensate without them	158
Connecting Boiler, Domes, etc	159
Expansion of Cast and Wrought Iron	161
A Table of Linear Expansion of Wrought- and Cast-iron Pipes	162

CHAPTER XV.

SIZE OF MAIN-PIPES FOR LOW-PRESSURE STEAM-HEATING.

Size of Mains	163
Loss of Heat from Imperfect Apparatus	164
Heat or Power necessary to put Water into Boilers	165
Poor Economy to use Small Piping	165
Necessity for providing for a Direct Return	166
How to determine the Size of the Main	168
The *Unit* of Size in Pipes	169
Relation between Heating Surface and Diameter of Pipe	169
Diagram of the Size of Main-pipes for Gravity Apparatus	170

CHAPTER XVI.

STEAM.

Temperature of Steam	174
Technical Terms	174
Table of Elastic Force, Temperature, and Volume of Steam	177
Calculations on Steam, Water, etc	178
Diagram of Rankine's Formula	180

CHAPTER XVII.

HEAT OF STEAM.

The Unit of Heat	182
Sensible and Latent Heat of Steam	182
A Diagram of Sensible and Latent Heat of Steam and Water	184
Equivalents of Heat	186

CHAPTER XVIII.

AIR.

	PAGE
What Air is	188
Air necessary for an Adult	189
Expansion of Air	191
Watery Vapor in the Atmosphere	193
Quantity of Moisture Air is capable of taking up	194
Drying Power of Air	194
A Table of the Watery Vapor Air is capable of taking up	195
What does Ventilation cost?	198

CHAPTER XIX.

HIGH-PRESSURE STEAM USED EXPANSIVELY IN PIPES FOR POWER AND HEATING.

Systems	200
New York Steam Co.'s System	202

CHAPTER XX.

EXHAUST STEAM AND ITS VALUE.

Thermal Value	215
How Hot can Feed-water be made	216
What Percentage of the Coal Heap does the Heating of the Feed-water represent	218
How Much of the Exhaust Steam can be used in warming the Feed-water	218
Warming Buildings with Exhaust Steam	218
Loss from Back Pressure	218
Exhaust and Live Steam in the Same Coils	222

CHAPTER XXI.

EXHAUST-STEAM HEATING.

Piping	223
Two Kinds of Steam used in Heating System	224

	PAGE
General Plan of an Office-building Plant	224
Feed-water Connections	226
Grease Separator	227
Check-valve	227
Reducing-pressure Valve	229
Receiving Tank and Pump Governor	231
Pumps	231
Place for Grease Separator	232

CHAPTER XXII.

THE SEPARATION OF GREASE FROM EXHAUST STEAM.

Effect of Grease in Boilers	233
Size an Important Feature	234
Method of Separation	234
Kind and Size of Tanks used	236
Advantage of Large Tanks to act as Mufflers on Heating System	236
Operation of the Baldwin Grease Separator	237
Combined Grease Separator and Feed-water Heater	241

CHAPTER XXIII.

BOILING AND COOKING BY STEAM, AND HINTS AS TO HOW THE APPARATUS SHOULD BE CONNECTED.

Steaming in the Atmosphere	242
Connections to Steamers	244
Kind of Pipe to be used	244
Vapor Pipe	245
Water Seal	245
Steaming under Pressure	246
Steam-kettles	247
Reason why Large Connections to Kettles should be used	251
Warming Water in Tanks	252
Warming Water for Bath and Laundry Purposes	254
Steam Roasting Ovens	257

INDEX. 363

CHAPTER XXIV.

DRYING BY DIRECT STEAM.

	PAGE
Description	259
Laundry-drying	261
Dry Kilns and Other Modes of Drying	266

CHAPTER XXV.

DRYING BY AIR CURRENTS.

Drying Bricks	269
Expense of moving Air with Fan	270
Moving Air by Means of an Aspirating Chimney (Fig. 113)	272
Why Goods appear Damp when removed from the Drying Room	273

CHAPTER XXVI.

STEAM TRAPS.

Object of Steam Trap	274
Two Classes	274
Direct-return Trap—Its Principle and Operation	275
Automatic Steam Trap	277
Open Float or Pot Trap	279
Modifications of the Pot Trap	282
Best Form of Automatic Steam Trap	284

CHAPTER XXVII.

VALVES FOR RADIATORS.

Kind of Valve to use	285
Radiator connected with Globe Valves showing Reason why they should not be used	286
Johnson Pneumatic System	289

CHAPTER XXVIII.

REMARKS ON BOILER CONNECTIONS AND ATTACHMENTS.

	PAGE
Feed Pipes	292
Check-valves	292
Blow-off Cocks	292
Safety-valves	294
Gauge or Try Cocks	295
Glass Water Gauges	295
Steam Gauges	296
Main Steam-pipe for Heating Apparatus	296

CHAPTER XXIX.

DATA ON CONDENSATION IN RADIATORS.

Results of Tredgold's Experiments	298
Results of Hood's Experiments	301
To calculate Heat Units radiated per Square Foot per Hour per Degree Difference of Temperature	304
Effect of Glass Windows in a Room	
Effect of Wind on Glass	305
Results of Barrus' Experiments	306
Results of the Writer's Experiments	307
Denton and Jacobus' Experiments with Extended Surfaces	309
Results of Tests made on some of the Regular Makes of Sectional Cast-iron Radiators by the Writer	311

CHAPTER XXX.

PIPE COVERING—WHAT IS SAVED THEREBY, AND OTHER DATA.

Analogy of Weight to the Efficiency	322
Relative Efficiency of the Different Makes	323
Results of Tests on Twelve Different Makes of Covering (Table No. 8)	324
Other Conditions than Weight to be considered in selecting a Covering	324

CHAPTER XXXI.

MISCELLANEOUS NOTES.

	PAGE
Cutting Recesses or Chases for Risers	325
Covering Recesses	326
Turning Exhaust Steam or Vapor into Chimneys	326
Soldering of Brass Feltings	328
Action of Solder under High Pressure	328
Frost-bursts	329
Painting Pipes	329

CHAPTER XXXII.

FIRE FROM STEAM-PIPES.

Danger from Superheated Steam	332
Results of Stahl's Experiments on Preparation of Charcoal	333
Results of Writer's Experiments	333

CHAPTER XXXIII.

MISCELLANEOUS NOTES AND TABLES.

Weight of a Cubic Inch of Various Metals (Table No. 9)	338
Weight of a Cubic Foot of Various Building Materials in Pounds (Table No. 10)	339
To find the Weight of Iron Castings by Computation	339
To find the Weight of Irregular Castings	339
Difference between American and English Wire Gauges and the Thickness of Plates in Decimals of an Inch (Table No. 11)	340
Definitions and Computations in Mensuration required by the Steam Fitter, etc., etc.	341
Diameters, Circumferences, and Areas (Table No. 12)	343
Number of Feet in Length of Various-sized Pipe which will contain One Cubic Foot of Water (Table No. 13)	347
Horizontal Multi-tubular Boilers	347
Boiler Data (Table No. 14)	353

ADVERTISEMENTS.

A CARD BY THE AUTHOR.

WM. J. BALDWIN,

Mem. Am. Soc. Mech. Eng.   Mem. Am. Soc. C. E.

HEATING AND VENTILATING
ENGINEER AND CONTRACTOR.

BUILDINGS WARMED BY
STEAM OR HOT WATER,
AND
VENTILATED BY THE MOST APPROVED METHODS.

AUTHOR OF

"STEAM-HEATING FOR BUILDINGS,"
(*Now in Fourteenth Edition.*)
THE "THERMUS" PAPERS,
"HOT-WATER HEATING AND FITTING,"
ETC., ETC.

CONSULTING ENGINEER IN WARMING AND VENTILATING.

106 & 108 B<small>EEKMAN</small> S<small>T</small>., N<small>EW</small> Y<small>ORK</small>.

THE JOHNSON SYSTEM

OF

TEMPERATURE REGULATION

FOR

PUBLIC AND OFFICE BUILDINGS, HOSPITALS, SANITARIUMS, SCHOOLS, AND RESIDENCES

Maintains a Uniform Temperature in All Rooms and Apartments regardless of External Changes.

NO ELECTRICITY. MOTIVE FORCE, COMPRESSED AIR. APPLICABLE TO ALL HEATING SYSTEMS.

Address for Descriptive Catalogue:

JOHNSON TEMPERATURE REGULATING CO.,

240 FOURTH AVENUE, NEW YORK CITY.

ALSO MANUFACTURERS OF

REDUCING VALVES, HIGH-PRESSURE GOVERNORS, AND HOT-WATER REGULATORS.

THE H. B. SMITH CO.,
133-135 CENTRE STREET, NEW YORK,

COTTAGE BOILER.

MANUFACTURERS
OF
HEATING
APPARATUS
FOR
Warming All Classes of Buildings with
STEAM
OR
WATER.

MILL'S SAFETY SECTIONAL BOILER,
ADAPTED FOR STEAM OR WATER.

MERCER, GOLD, and COTTAGE
BOILERS,
Arranged for Hard or Soft Coal and Wood.

UNION, ROYAL UNION, CHAMPION, CORONET, and CROWN
RADIATORS.

GOLD'S PIN INDIRECT RADIATORS.

New York, Providence, Philadelphia.
Foundry: WESTFIELD, MASS.

SEND FOR CIRCULAR.

ROYAL UNION RADIATOR.

KIELEY'S STEAM SPECIALTIES.

Pump Governors,
Steam Traps,
Steam and Water Reducing Valves,
Water Arches,
Noiseless Back-Pressure Valves,

Oil Separators, Temperature Controllers for Hot-Water Tanks,
Damper Regulators and Regulating Valves for Elevator, Fire, and Roof Tank Pumps.

KIELEY & MUELLER,

7 to 17 West Thirteenth Street, NEW YORK.

KIELEY'S WATER=SEAL SYSTEM OF STEAM HEATING
AND AUTOMATIC RETURN OF CONDENSATION.

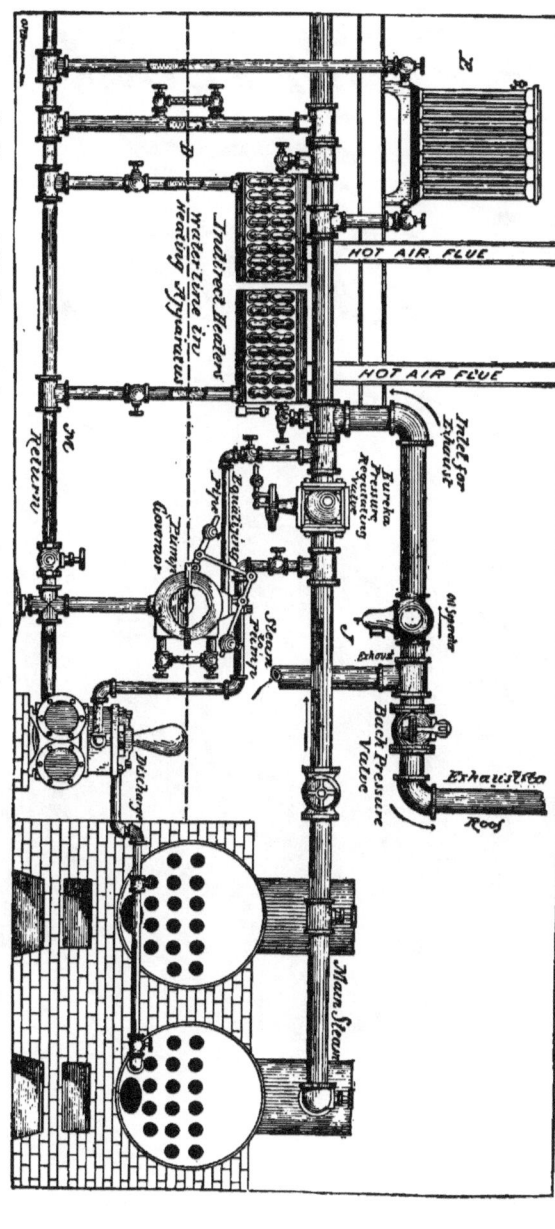

By the use of this System a large saving of Fuel and a Noiseless Working Apparatus is guaranteed. The specialties mentioned on opposite page are required to obtain this system.

KIELEY & MUELLER, 7 to 11 West 13th Street, New York.

v

Gurney Boilers

For Hot Water.   **For Steam.**

"300 Series."   "Doric."
"Doric."   "Gurney."
"Double Crown."   "Bright Idea."
"Bright Idea."   "Defiance."
"Defiance."

They are the product of long experience and the highest possible intelligence in the study of heating.

**ENDORSED BY THE LEADING ENGINEERS.
SEND FOR LATEST TRADE
CATALOGUE AND PRICE LISTS.**

GURNEY HEATER MFG. CO.

Home Office, 163 Franklin Street, BOSTON, MASS.
New York Branch, 48 Centre Street, NEW YORK, N. Y.

THE CALDWELL WATER-TUBE BOILER,

Maximum Evaporation. Minimum Cost of Maintenance.

We Baffle the Gases—Not the Water.

EFFICIENCY,
SAFETY,
SIMPLICITY,
ECONOMY,
STRENGTH,
ACCESSIBILITY,
DURABILITY.

All of which we can demonstrate to your entire satisfaction.

QUOTATIONS FROM TESTIMONIALS.

"Beyond our expectations."
"About 15% less expenditure for coal."
"We made no mistake."
"Have shown no sign of a leak."
"Simply no trouble whatever."
"In two years' continuous operation we have not used a hammer or wrench on them."

Write for Catalogue, Tests, and Testimonials.

We court the fullest inquiry.
Correspondence Solicited.

JAMES BEGGS & CO.,
MANUFACTURERS,
No. 9 DEY STREET, NEW YORK.

The Acton Automatic Steam and Water Specialties,
79 & 81 Washington St., Brooklyn, N. Y.

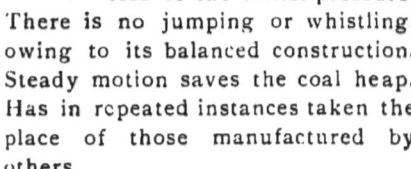

FIG. 0.

FIG. 0 represents view of the **ACTON LEATHER=CUP GOVERNOR**. It is particularly adapted for filling house-tanks and controlling the pumps of hydraulic elevators. It is a double-seated valve, absolutely tight.

FIG. 1. For steam or water. **PRESSURE=REGULATOR** will control pressure at atmosphere or any desired variation, regardless of the initial pressure. There is no jumping or whistling, owing to its balanced construction. Steady motion saves the coal heap. Has in repeated instances taken the place of those manufactured by others.

ACTON NOISELESS BACK-PRESSURE VALVE will work under any and all conditions. I claim

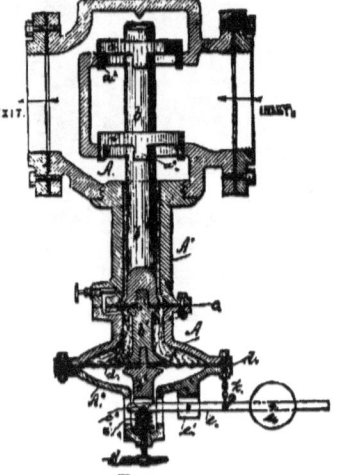

FIG. 1.

this valve second to none. It is absolutely tight and free under all conditions. Made in sizes from 2 to 24 inches.

The **ACTON AUTOMATIC PUMP = RECEIVER AND REGULATOR** has many features which assists the water of condensation from buildings or heating coils to flow to the pump easily and readily; which latter returns it to the boilers.

ALSO MANUFACTURER OF
REDUCING AND RELIEF VALVES FOR WATER, DAMPER REGULATORS, BALANCED VALVE FOR FLOAT PURPOSES.

All goods manufactured by us are guaranteed satisfactory.

L. J. WING & COMPANY,

109 Liberty St.,

NEW YORK CITY,

MANUFACTURERS OF AND DEALERS IN

WING'S DISC FANS
AND
COMBINATIONS,

Stationary, Marine, and Motor=Wagon

GAS ENGINES.

WING'S DISC FANS are the best made in this or any other country, as is fully shown by two plain facts, viz., they have taken the highest premiums wherever exhibited in competition all over the world, and there are more in use than all other fans put together.

For heating, ventilating, cooling, drying, and the hundreds of uses in factories, mines, steamships, etc., they lead in efficiency and power used for the amount of air moved.

They are built for running by belt, also with high-speed steam-engine or electric-motor combination.

WING'S GAS ENGINES are of new design and principle, very compact, simple, and reliable. They are very valuable for pumping, elevators, electric-light plants, etc. They will run on any gas, or make their own gas automatically from 76-degree naphtha or gasoline. They are used much in marine service.

SEND FOR CATALOGUE.

If you want to heat water with steam write for our Wainwright catalogue. Have you seen our Expansion Joint with Equalizing Rings?

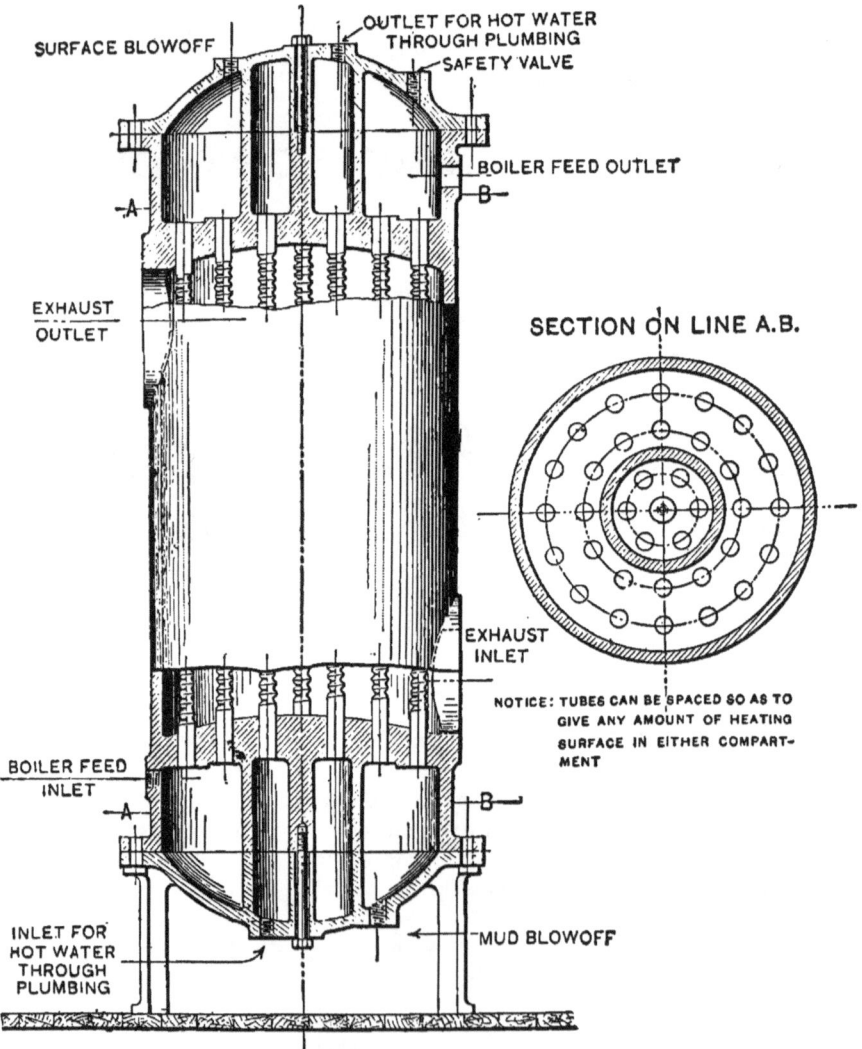

WAINWRIGHT
COMBINATION HEATER
AS BUILT BY
TAUNTON LOCOMOTIVE MFG. CO.,
TAUNTON, MASS.

BALDWIN'S GREASE SEPARATOR

(PATENTED)

Is guaranteed to take ALL grease out of exhaust steam.

The apparatus can be seen in use in the following buildings:

THE HANOVER FIRE INSURANCE CO.,
34 Pine St., New York.

THE MANHATTAN AND MERCHANTS' BANK,
40 & 42 Wall St., New York.

THE AMERICAN THEATRE,
42d St. & 8th Ave., New York.

THE COLLEGE OF PHYSICIANS AND SURGEONS,
59th St., 9th & 10th Aves., New York.

THE MANHATTAN EYE AND EAR HOSPITAL,
41st St. & 4th Ave., New York.

THE PHELPS MEMORIAL HALL,
Yale University, New Haven.

TRINITY SCHOOL,
W. 90th St., New York.

THE NEW YORK TRIBUNE BUILDING.

THE MECHANICS' BANK BUILDING,
Montague St., Brooklyn, N. Y.

And many other buildings.

FOR CATALOGUE AND PARTICULARS ADDRESS

WM. J. BALDWIN, 106-108 Beekman St., New York.

METROPOLITAN HIGH AND LOW PRESSURE SELF-CONTAINED SIDE-CRANK ENGINE.

• • •

CONSTRUCTED ON MECHANICAL PRINCIPLES THROUGHOUT.

• • •

MADE FOR SCHOOLS AND PUBLIC BUILDINGS.

• • •

DONEGAN & SWIFT, 6 Murray Street, New York.

JENKINS BROS.' VALVES

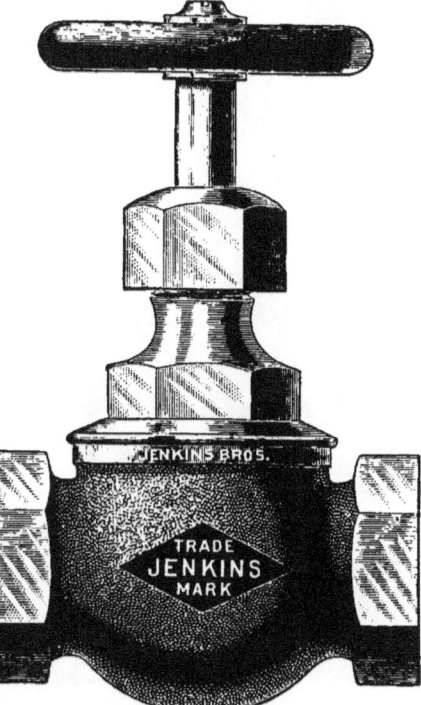

Have Keyed Stuffing-box attachment and Removable Disc Nut. Warranted full opening.

Should you order, insist on having valve stamped like cut with our Trade-mark.

THE JENKINS AUTOMATIC AIR VALVE

Has proved to be all we claimed for it, *i.e.*, A Positive and Reliable Air Valve. The expansible plug is a special compound and manufactured expressly for use in the Jenkins Automatic Air Valve. It does not deteriorate or lose its flexibility under the action of heat or steam.

With the Jenkins Automatic Air Valve you should connect with a drip pipe, thus insuring a positive circulation.

JENKINS BROS., New York, Boston, Chicago, Philadelphia.

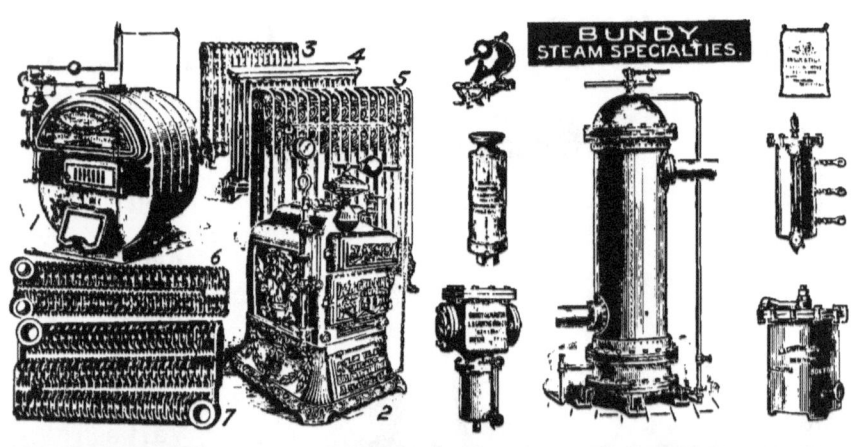

66-68 CENTRE STREET, NEW YORK.

BOSTON: **PHILADELPHIA:**
177-179 Fort Hill Square.   702 Arch Street.

Works: JERSEY CITY, New Jersey, U. S. A.

HEADQUARTER FOR

Bundy Heater,	Bundy Feed Water Heater,
Bundy "La Villa" Heater,	Bundy Steam Trap,
Bundy "Columbia" Radiator,	Bundy Exhaust Head,
Bundy "Standard" Radiator,	Bundy Steam and Oil Separator,
Bundy "Climax" Radiator,	Bundy Insulation Covering,
Bundy "Elite" Radiator,	Bundy Low Water Alarm,
Bundy "Newport" Radiator,	Bundy Hydrox Steam Trap.

SEND FOR BOOK B. B.

CLEVAUC
TRADE MARK

STEAM SPECIALTIES.

"CLEVAUC"
BACK-PRESSURE VALVE.

"CLEVAUC"
PRESSURE REGULATOR.

NOISELESS BACK-PRESSURE VALVES.
  REDUCING-PRESSURE VALVES, ANY STYLE.
    AUTOMATIC AIR VALVES.
      ELEVATOR-PUMP GOVERNORS.
        AIR-COMPRESSOR GOVERNORS.
          TEMPERATURE REGULATORS FOR HOT-WATER TANKS.
            RECEIVING TANK AND GOVERNORS COMBINED.
              AUTOMATIC DAMPER REGULATORS.
                AUTOMATIC RETURN STEAM TRAPS.
                  SAFETY WATER COLUMNS.
                    AUTOMATIC WATER FEEDERS.
                      STEAM TRAPS.
                        HOUSE-PUMP GOVERNORS.

We control all Van Auken Patents issued since 1893.

ALL GOODS SENT ON 30 DAYS' TRIAL TO RESPONSIBLE PARTIES.

The Clarence E. Van Auken Co.

Home Office and Works:
166-174 S. CLINTON STREET,
CHICAGO,

New York House:
125-127 WORTH STREET,
NEW YORK CITY.

www.ingramcontent.com/pod-product-compliance
Lightning Source LLC
Chambersburg PA
CBHW022119290426
44112CB00008B/736